AF824038

Reports of China's Basic Research

Editor-in-Chief

Wei Yang, National Natural Science Foundation of China, Beijing, China, Zhejiang University, Hangzhou, Zhejiang, China

The National Natural Science Foundation of China (NSFC) was established on February 14, 1986. Upon its establishment, NSFC was an institution directly under the jurisdiction of the State Council, tasked with the administration of the National Natural Science Fund from the Central Government. In 2018, it became managed by the Ministry of Science and Technology (MOST) but kept its due independence in operation. Since its establishment, NSFC has comprehensively introduced and implemented a rigorous and objective merit-review system to fulfill its mission of supporting basic research, fostering talented researchers, developing international cooperation and promoting socioeconomic development.

Featuring science, basics, and advances, the series of *Reports of China's Basic Research* is organized by the NSFC to present the overall level and pattern of China's basic research, share innovative achievements, and illustrate excellent breakthroughs in key fields. It covers various disciplines including but not limited to, computer science, materials science, life sciences, engineering, environmental sciences, mathematics, and physics. The series will show the core contents of the final reports of the Major Programs and the Major Research Plans funded by NSFC, and will closely follow the frontiers of basic research developments in China.

If you are interested in publishing your book in the series, please contact Qian Xu (Email: violetta_xuqian@zju.edu.cn) and Mengchu Huang (Email: mengchu.huang@cn.springer.com).

国家自然科学基金委员会
1986
NATIONAL NATURAL SCIENCE FOUNDATION OF CHINA

Rixiang Zhu
Editor

Research on Destruction of the North China Craton

ZHEJIANG UNIVERSITY PRESS
浙江大学出版社

Springer

Editor
Rixiang Zhu
Institute of Geology and Geophysics
Chinese Academy of Sciences
Beijing, China

ISSN 2731-8907 ISSN 2731-8915 (electronic)
Reports of China's Basic Research
ISBN 978-981-99-6045-3 ISBN 978-981-99-6046-0 (eBook)
https://doi.org/10.1007/978-981-99-6046-0

Jointly published with Zhejiang University Press
The print edition is not for sale in China (Mainland). Customers from China (Mainland) please order the print book from: Zhejiang University Press.
ISBN of the co-publisher's edition: 978-7-308-22762-9

© Zhejiang University Press 2023

This work is subject to copyright. All rights are solely and exclusively licensed by the Publisher, whether the whole or part of the material is concerned, specifically the rights of translation, reprinting, reuse of illustrations, recitation, broadcasting, reproduction on microfilms or in any other physical way, and transmission or information storage and retrieval, electronic adaptation, computer software, or by similar or dissimilar methodology now known or hereafter developed.
The use of general descriptive names, registered names, trademarks, service marks, etc. in this publication does not imply, even in the absence of a specific statement, that such names are exempt from the relevant protective laws and regulations and therefore free for general use.
The publishers, the authors, and the editors are safe to assume that the advice and information in this book are believed to be true and accurate at the date of publication. Neither the publishers nor the authors or the editors give a warranty, expressed or implied, with respect to the material contained herein or for any errors or omissions that may have been made. The publishers remain neutral with regard to jurisdictional claims in published maps and institutional affiliations.

This Springer imprint is published by the registered company Springer Nature Singapore Pte Ltd.
The registered company address is: 152 Beach Road, #21-01/04 Gateway East, Singapore 189721, Singapore

Paper in this product is recyclable.

Editorial Board

Editor-in-Chief

Wei Yang

Associate Editors

Ruiping Gao
Yu Han

Editors

Changrui Wang
Qidong Wang
Xuelian Feng
Liexun Yang
Junlin Yang
Liyao Zou
Zhaotian Zhang
Xiangping Zhang
Yongjun Chen
Yanze Zhou
Ruijuan Sun
Jianquan Guo

Longhua Tang
Guoxuan Dong
Zhiyong Han
Ming Li

Preface to the Series

As Lao Tzu said, "A huge tree grows from a tiny seedling; a nine-storied tower rises from a heap of earth." Basic research is the fundamental approach to fostering innovation-driven development, and its level becomes an important yardstick for measuring the overall scientific and national strength of a country. Since the beginning of the twenty-first century, China's overall strength in basic research has been consistently increasing. With respect to input and output, China's input in basic research increased by 14.8 times from 5.22 billion yuan in 2001 to 82.29 billion yuan in 2016, with an average annual increase of 20.2%. In the same period, the number of China's scientific papers included in the Science Citation Index (SCI) increased from lower than 40,000 to 324,000; China rose from the sixth to the second place in global ranking in terms of the number of published papers. In regard to the quality of output, in 2016, China ranked No. 2 in the world in terms of citations in nine disciplines, among which the materials science ranked No. 1; as of October 2017, China ranked No. 3 in the world in the numbers of both Highly Cited Papers (top 1%) and Hot Papers (top 0.1%), with the latter accounting for 25.1% of the global total. In talent cultivation, in 2006, China had 175 scientists (136 of whom from the Chinese mainland) included in Thomson Reuters' list of Highly Cited Researchers, ranking fourth globally and first in Asia.

Meanwhile, we should also be keenly aware that China's basic research is still facing great challenges. First, funding for basic research in China is still far less than that in developed countries—only about 5% of the R&D funds in China are used for basic research, a much lower percentage than 15%–20% in developed countries. Second, competence for original innovation in China is insufficient. Major original scientific achievements that have a global impact are still rare. Most of the scientific research projects are just a follow-up or imitation of existing research, rather than groundbreaking research. Third, the development of disciplines is not balanced, and China's research level in some disciplines is noticeably lower than the international level—China's Field-Weighted Citation Impact (FWCI) in disciplines just reached 0.94 in 2016, lower than the world average of 1.0.

The Chinese government attaches great importance to basic research. In the 13th Five-Year Plan (2016–2020), China has established scientific and technological innovation as a priority in all-round innovation and has made strategic arrangements to strengthen basic research. General Secretary XI Jinping put forward a grand blueprint for making China a world-leading power in science and technology in his speech delivered at the National Conference on Scientific and Technological Innovation in 2016 and emphasized that "we should aim for the frontiers of science and technology, strengthen basic research, and make major breakthroughs in pioneering basic research and groundbreaking and original innovations" at the 19th CPC National Congress on October 18, 2017. With more than 30 years of unremitting exploration, the National Natural Science Foundation of China (NSFC), one of the main channels for supporting basic research in China, has gradually shaped a funding pattern covering research, talent, tools, and convergence, and has taken action to vigorously promote basic frontier research and the growth of scientific research talent, reinforce the building of innovative research teams, deepen regional cooperation and exchanges, and push forward multidisciplinary convergence. As of 2016, nearly 70% of China's published scientific papers were funded by the NSFC, accounting for 1/9 of the total number of published papers all over the world. Facing the new strategic target of building China into a strong country in science and technology, the NSFC will conscientiously reinforce forward-looking planning and enhance the efficiency of evaluation, so as to achieve the strategic goal of making China progressively share the same level with major innovative countries in research total volume, contribution and groundbreaking researchers by 2050.

The series of *Advances in China's Basic Research* and the series of *Reports of China's Basic Research* proposed and planned by the NSFC emerge against such a background. Featuring science, basics, and advances, the two series are aimed at sharing innovative achievements, diffusing performances of basic research, and leading breakthroughs in key fields. They closely follow the frontiers of basic research developments in China and publish excellent innovation achievements funded by the NSFC. The series of *Advances in China's Basic Research* mainly presents the important original achievements of the programs funded by the NSFC and demonstrates the breakthroughs and forward guidance in key research fields; the series of *Reports of China's Basic Research* shows the core contents of the final reports of Major Programs and Major Research Plans funded by the NSFC to make a systematic summarization and give a strategic outlook on the achievements in the funding priorities of the NSFC. We hope not only to comprehensively and systematically introduce backgrounds, scientific significance, discipline layouts, frontier breakthroughs of the programs, and a strategic outlook for the subsequent research, but also to summarize innovative ideas, enhance multidisciplinary convergence, foster the continuous development of research in concerned fields, and promote original discoveries.

As Hsun Tzu remarked, "When earth piles up into a mountain, wind and rain will originate thereof. When waters accumulate into a deep pool, dragons will come to live in it." The series of *Advances in China's Basic Research* and *Reports of China's Basic Research* are expected to become the "historical records" of China's basic research. They will provide researchers with abundant scientific research material and vitality of innovation, and will certainly play an active role in making China's basic research prosper and building China's strength in science and technology.

Wei Yang
Academician of the Chinese Academy of Sciences
Beijing, China

The plan has achieved a number of research results with major innovations as follows: accumulating important basic data for studying continental evolution; developing a series of *in situ* microchronology, isotope analysis methods and deep exploration techniques; revealing the deep textures and structures of the NCC; determining the temporal and spatial differences of Mesozoic to Cenozoic magmatism; constraining the temporal and spatial range of the NCC destruction; confirming that the westward subduction of the Pacific Plate in the Early Cretaceous was the first-order external control factor and driving force for the NCC destruction; exploring the shallow effect of craton destruction; confirming that the formation of gold deposits in the Jiaodong area was mainly controlled by large-scale magmatism during the NCC destruction; proposing that the essence of craton destruction is that the material composition and physicochemical properties of the lithospheric mantle have changed fundamentally, so the inherent stability of the craton has been destroyed; and establishing the theoretical system of craton destruction. These solid earth science research achievements in China have exerted the most international influence in recent years and played a leading role for Chinese geoscientists to take in the world.

As part of the book series *Reports of China's Basic Research*, this book summarizes the scientific ideas, key scientific issues, scientific objectives, and models of the Plan "Destruction of the North China Craton", demonstrates its important achievements and significant progress, and makes a strategic prospect for the study of global dynamics.

Beijing, China   Rixiang Zhu

Contents

Contributors

Rixiang Zhu Institute of Geology and Geophysics, Chinese Academy of Sciences, Beijing, China

Guowei Zhang Northwestern University, Xi'an, China

Zhenmin Jin China University of Geosciences, Wuhan, China

Shuguang Li China University of Geosciences, Wuhan, China

Guangtian Zou Jilin University, Changchun, China

Xiankang Zhang Geophysical Exploration Center, China Earthquake Administration, Zhengzhou, China

Weiming Fan Guangzhou Institute of Geochemistry, Chinese Academy of Sciences, Guangzhou, China

Project Overview

1 Introduction

The earth's surface consists of continents and oceans. Continents cover nearly one-third of the earth's surface and are essential to human survival. More than 95% of the natural resources used by human beings come from continents. Therefore, the formation and evolution of continents have always been the most fundamental scientific issue of solid earth sciences, and it is also the key to solving the problems of national resources and energy demand. People have conducted a long-term exploration in the formation, evolution, internal structure, and control over resources of continents. The theory of plate tectonics proposed in the 1960s has greatly promoted the development of geoscience. It explains the generation, development and demise of ocean plates, reveals the causes of earthquakes, magmatism and crustal deformation on the continental margin, demonstrates the seafloor spreading and horizontal plate movement, and summarizes the development laws of plate breakup, divergence, drift and convergence. Therefore, it has become a guiding theory to understand the development of global tectonics.

The theory of plate tectonics was founded on the in-depth understanding of ocean geology and tectonics. The theory holds that the lithospheric plate is rigid, and the crustal deformation and magmatism are limited to the margin of the plate. This understanding is more applicable to the oceanic lithosphere. The research results of continental geology show that the continental crust has widely undergone tectonic deformation, and various degrees of magmatism and metamorphism have also occurred. However, the theory of plate tectonics cannot provide a reasonable explanation for continental geological processes and dynamic mechanisms. Further research shows that there are great differences between continental and oceanic lithospheres in material composition, structure, rheology, and deep thermal structure. The understanding of the uniqueness of continental evolution prompted us to re-examine continental dynamics and seek a theory that can reasonably explain continental evolution. Therefore, the study of continental evolution and its dynamics provides a window to a new theory in geoscience. In recent years, a series of major research programs have been

© Zhejiang University Press 2023

R. Zhu (ed.), *Research on Destruction of the North China Craton*, Reports of China's Basic Research, https://doi.org/10.1007/978-981-99-6046-0_1

set up to explore the formation and evolution of continents and to establish a new theoretical system of continental evolution.

The continent is composed of active orogenic belts and stable cratons. This view point was first proposed in the geosyncline-platform theory, which is basically accepted and inherited by the plate tectonic theory. The stability of the craton results from low density, low water content and thick lithospheric root, so it can float above the asthenosphere. The thick and dry lithospheric mantle of the craton can resist reworking by later geological processes to a great extent. Therefore, the traditional theory holds that the craton is stable. Except for a few magmatic activities derived from the deep earth, the craton basically has no tectonic deformation in the lithosphere or crust and large-scale magmatism, and there is no strong seismic activity.

The NCC has preserved an ancient continental crust of more than 3.8 billion years. It entered a stable cratonic stage 1.8 billion years ago and was in a stable state for a long time before the Mesozoic. However, obvious lithospheric thinning and craton destruction occurred in the eastern part of the NCC in the Mesozoic, resulting in the disappearance of its original stability characteristics. This shows that cratons can be either long-term stable or unstable, which is a major geological phenomenon that cannot be explained by the classical plate tectonic theory. This raises an important scientific question: why can the originally stable craton be destroyed? The activity of the NCC from the Mesozoic to the Cenozoic has long been recognized by Chinese geologists who have different explanations. Early in the twentieth century, WENG Wenhao put forward the concept of "Yanshan Movement" according to the Jurassic–Cretaceous tectonic and magmatic (volcanic) activities in the north of the NCC. Subsequently, CHEN Guoda put forward the viewpoint of "rejuvenation of platform". In the 1990s, based on the research of previous data, Chinese and foreign scholars found that the thickness of the lithosphere in the eastern part of the NCC has been thinned by more than 100 km, so they put forward the concept of "lithospheric thinning" or "unrooting".

Although many geoscientists have studied the NCC for a long time, the temporal and spatial scope and dynamic mechanism of the loss of stability of the NCC have been perplexing geoscientists. Why the craton lost stability has been unresolved. Therefore, why the craton was destroyed has been a puzzle for geoscientists for nearly a century.

The NCC is a typical example of global craton destruction. This is China's regional advantage, which is conducive to the field research. The National Natural Science Foundation of China launched the major research plan "Destruction of the North China Craton" in 2007, which ended at the end of 2017. According to the needs of completing the scientific objectives, the Plan has set up 66 discipline and comprehensive integration projects, with craton destruction as the core scientific issue, based on the original data obtained from observations and experiments and with the layout of geological tectonics, geophysics, petrology and geochemistry, experimental simulation, resources and environmental disasters, discipline integration, and strategic research. The total funding reached 200 million yuan, involving the Department of Earth Sciences, the Department of Mathematical and Physical Sciences, the Department of Information Sciences and other academic departments.

The Plan concentrates China's advantageous research forces in the fields of earth sciences, mathematical sciences and information sciences, breaks through the constraints of traditional disciplines, and carries out cross integration among different disciplines in an effective way. At the same time, the Plan has made breakthroughs on the destruction and reworking after the formation of the NCC and on its tectonic, magmatic, metallogenic and paleoenvironmental effects so that China can make original contributions to improving the theoretical system of earth formation and evolution and establishing the theoretical framework of earth sciences.

After 10 years of efforts, the Plan makes use of the comprehensive advantages of geology, geophysics, geochemistry and other disciplines, and understands the spatio-temporal distribution and processes of the NCC destruction from the perspective of earth system sciences through observations, experiments and theoretical analyses. The Plan reveals the interaction between the lithosphere and asthenosphere and the dynamic mechanism of the interior craton destruction, establishes the theoretical system of craton destruction, explores the resource effect of craton destruction, and improves the understanding of continent formation and evolution.

Through the implementation of the Plan, "the North China Craton Destruction" has become a hotspot in the study of global continental evolution and dynamics; the leading role of Chinese scientists in this field has been brought into play; and the international academic status of China's solid earth science research has been significantly improved.

2 Research Circumstances

2.1 Overall Scientific Objectives

By highly integrating the analytical technology and detection means of modern science, the Plan aims to achieve the following overall scientific objectives, including obtaining the observation, experiment and theoretical research results led by high and new technology; understanding the spatio-temporal distribution, processes and mechanism of the NCC destruction; revealing the properties, structure and interaction of materials in the interior of craton destruction; clarifying the shallow effect of craton destruction and its implications for mineral resources, energy and disasters; and improving the understanding of the formation and evolution of continents.

2.2 *Key Scientific Issues*

The Plan intends to solve seven key scientific issues:

(1) Spatio-temporal distribution of the NCC destruction;
(2) The relationship between the deep process of the NCC destruction and global events;
(3) Shallow effect of the NCC destruction;
(4) The relationship between the NCC destruction and the accumulation of mineral resources;
(5) The relationship between the NCC destruction and the current seismic activity;
(6) The mechanism, process and dynamics of the NCC destruction;
(7) Significance of craton destruction in global geology and continental evolution.

3 Research Work

The main scientific research work completed in the Plan can be summarized as follows.

1. **Multidisciplinary whole-scale observation and research**

In the 10 years since the implementation of the Plan, 33 projects on geology, geophysics and petrochemistry have continued to carry out geological investigation, rock sampling, geochemical experiments and deep structure exploration in the NCC and its adjacent areas to obtain the original information as fully as possible. Relying on advanced detection and experimental technology, the original samples and data are analyzed and studied with high precision (resolution). The research basically achieves multidisciplinary observation covering the target area.

(1) Geological investigation

The project has investigated and studied the deformation along the Tan–Lu Fault Zone, metamorphic core complexes and extension domes in Jiaodong, Liaodong and Inner Mongolia, extension and shortening deformation of the Yanshan and Taihang Mountains, sediments and basins in the Yanshan Tectonic Belt, nappe structures around Ordos, magmatism and tectonic deformation in eastern China.

(2) Rock sampling and geochemical experiments

Field investigation and rock sampling were carried out on the products of typical magmatic activities of the Paleozoic, Mesozoic and Cenozoic in different areas and deep-seated xenoliths and mineral xenocrysts. The research samples mainly include Mesozoic and Cenozoic basalts and mantle xenoliths, Mesozoic mafic magmas, Mesozoic granites and adakites, the terrane granulite and the lower crustal granulite xenoliths entrained in the volcanic rocks from different periods. The researchers have conducted a systematic study on the geochronological and geochemical analysis of a large number of rock samples exposed in the eastern part of the craton, such as the Liaodong, Shandong, Hannuoba and Linxi areas, Yanshan Tectonic Belt, Korean Peninsula, Tan–Lu Fault Zone, Central Orogenic Belt, and Dabie–Sulu Orogenic Belt.

(3) Seismic exploration of deep structure

To detect the lower crust and upper mantle structures of the NCC and its adjacent areas, the Plan has been fully deployed and implemented for seismic exploration covering the whole region. Eight sections and one two-dimensional seismic array with a total of 688 stations were deployed in and around the NCC. Combined with the complementarity of active source seismic detection, three wide-angle reflection/refraction profiles with long-range artificial seismic source (3650 km) and two seismic observation profiles using the combined ocean-bottom-seismometer (OBS) and land portable seismometer (860 km) were completed. Through the development and application of effective seismic imaging technology, information on the seismic wave velocity distribution, the intermittent velocity interface structure of the crust and the upper mantle, and the anisotropy of the upper mantle in North China was obtained from seismic data. It lays a foundation for understanding craton destruction and exploring the formation and evolution mechanism of continents from the dynamic processes in the earth's interior.

2. **Study on the nature and evolution law of craton destruction**

The Plan has comprehensively carried out research on the nature and evolution law of craton destruction, and the main research results are as follows:

(1) The time range of the NCC destruction is clearly defined.
(2) The spatial range of the NCC destruction is clearly defined.
(3) The surface geological response to the NCC destruction is identified.
(4) The transformation of the crust caused by the NCC destruction is recognized.
(5) The changes in lithospheric composition and properties and their mechanisms are found and studied.
(6) The rich water characteristics of the Mesozoic lithospheric mantle in North China are identified.
(7) The influence of Paleo-Pacific Plate subduction on the structure and state of the upper mantle in North China is detected.
(8) It is proposed and demonstrated that the late Mesozoic subduction and retreat of the western Pacific Plate and the corresponding mantle action are the dynamic mechanisms leading to the destruction of the NCC.
(9) Based on the above understanding of the regularity of the evolution of the NCC, a new concept of "Craton Destruction" is proposed: "The inherent stability of the craton is destroyed due to the fundamental transformation of the material composition and physicochemical properties of the lithospheric mantle." The theory of craton destruction is established, which reveals the important position of the NCC destruction in global continental evolution and develops the theory of plate tectonics.

3. **Preliminary exploration on the resources and biological effects of craton destruction**

Based on the understanding of the nature and law of the NCC destruction, the Plan has carried out scale detection and obtained the following basic cognition about the resource effects of craton destruction: The formation of gold deposits in the Jiaodong area was mainly controlled by large-scale magmatism during the destruction of the NCC; the NCC destruction was conducive to the formation of large petroliferous basins in the eastern NCC; the NCC destruction had different effects on the preservation of pre-Mesozoic oil and gas in the eastern and western NCC.

Based on the comparison of vertebrate diversity in the Yanliao and Jehol biotas and the analysis of the environmental background, it is preliminarily proposed that the destruction of the NCC controlled the two biotas, and in practice, the two biotas were affected by two episodes of the Yanshan Movement.

4. **Accumulation and sharing of basic data**

(1) Geological discipline integration

The Plan focuses on the geological correlation between different units of the NCC and the comprehensive study of the spatio-temporal variation law and dynamic mechanism of important geological events. At the same time, a series of key geological maps covering basement structures, Meso-Neoproterozoic and Paleozoic characteristic structures, Meso-Cenozoic structures of each period, etc. have been compiled.

(2) Geochemistry discipline integration

The geochemical evidence of the destruction of the NCC has been summarized. The compiled drawings include the characteristics and composition of the Mesozoic and Cenozoic lithospheric mantle, the age distribution of mantle peridotite xenoliths and lower crust xenoliths, and the distribution of alkaline rock belts.

(3) Geophysical discipline integration

The Plan has built the database and application system of the three-dimensional velocity structure data model "Crust and Upper Mantle Velocity Model of North China". The content of the model includes the data and images of buried depth and layered velocity of various velocity interfaces of the crust, the lithosphere and upper mantle, the data and images of velocity disturbance of the upper mantle, and the crustal velocity structure of typical sections. The system is released and shared with others.

These basic maps, structural models and corresponding databases provide basic data for further study of China's continental evolution.

Research Status and Development Trends

1 Continental Evolution and Dynamics

Because plate tectonics cannot fully explain the characteristic geological process and dynamic mechanism within the continent, scientists have been seeking a theory that can reasonably explain continental evolution. To reveal the structures and geological evolution of the continental lithosphere, many countries have implemented a series of major scientific plans. The National Science Foundation, Geological Survey and Department of Energy of the United States, jointly put forward the "Continental Dynamics Plan" for a period of 30 years (1990–2020). The National Science Foundation of the United States began to implement the earth lens program in 2004. The program took modern geophysics, drilling, remote sensing and information technology as the leading system and accurately described the structures and evolution of the North American continent to establish a regional geodynamic evolution model. The North Atlantic Scientific Commission organized and implemented the "European Continental Exploration" research program (1997–2003) in Europe, focusing on the evolution history of the continental lithosphere. Some important progress has been made in the abovementioned continental tectonic research plan of the international geoscience community, including the continental rheological state, intracontinental orogeny, genesis of magmatic activity, the relationship between the deep thermal structure and the surface system, and impact of continental evolution on the life process.

From the 1970s to the 1980s, China's geoscience research mainly tracked the development of international geosciences and interpreted the tectonic history of the Chinese mainland based on the theory of plate tectonics. Since the beginning of the twenty-first century, China's continental evolution research has gradually broken through the simple thinking mode of plate tectonics and carried out a series of original explorations. Supported by the National Natural Science Foundation of China, the Ministry of Science and Technology, and the Ministry of Education, a number of major scientific research programs on continental tectonics have been proposed and implemented, such as the "Continental Deep Subduction" project in the 973 Program

© Zhejiang University Press 2023

R. Zhu (ed.), *Research on Destruction of the North China Craton*, Reports of China's Basic Research, https://doi.org/10.1007/978-981-99-6046-0_2

to explore how the continental crust subducts to the deep part of the lithosphere and the formation and exhumation of ultrahigh-pressure rocks; the "Continental Mantle Plume" project in the 973 Program to study the dynamic process of the uplifting of deep continental mantle materials, the formation of large igneous provinces, and their impact on the extinction of surface organisms; and the major project of the NSFC "Pangea Gathering and Dispersion of East Asian Continent" to explore the distribution, convergence and divergence process of different continental blocks over a long geological history.

The long-term research accumulation in the Chinese mainland has provided a solid foundation for Chinese scholars to create new theories of continental evolution. The Chinese mainland, richly endowed by natural geological conditions, also provides a rare natural laboratory for China's geoscience circles to achieve theoretical breakthroughs. China's growing economic strength and increasing investment in scientific research have created good opportunities for collaborative innovation research, joint scientific and engineering research and breakthroughs in new theories of continental evolution.

2 Craton Destruction

The earth's surface consists of continents and oceans. Among them, the continent is composed of cratons (ancient continents) and orogenic belts (young continents). The former is stable, and the latter is active. This view was first put forward by the geosyncline-platform theory, then inherited in modern plate tectonics theory, and finally became the long-term leading thought in the geoscience circle.

Cratons refer to ancient and stable continental blocks, which have three remarkable characteristics: i. The formation age is old, and most of them have Archean age; ii. the lithosphere is 200 km thick; iii. the tectonics are stable without large-scale tectono-magmatic activity and large earthquakes. The craton has the most complete geological history (approximately 4.4 billion years) and is an ideal place to study and understand the evolution of the continent. The cratons distributing on continents on a large-scale are different from orogenic belts for their long-term stability.

Most of the cratons in the world have remained stable for a long time, and a few have been partially destroyed. For this reason, developed countries have neglected to conduct in-depth and systematic research on craton destruction for a long time.

Compared with other cratons in the world, the destruction degree of the NCC is the strongest, and the destruction phenomenon is the most typical. Therefore, the NCC is a typical example of the destruction of the global cratons. The NCC destruction is a major geological event that has led North China to become an important base for China's important energy (oil, gas, coal) and metal minerals (gold, iron, molybdenum, etc.). Craton destruction belongs to the category of continental evolution, which is an important frontier field of geoscience, and many aspects cannot be explained by classical plate tectonics theory. The NCC is a great area for studying continental evolution. It provides a good opportunity for Chinese earth scientists to develop plate tectonics and create a new continental evolution theory.

3 Field Development Trends

The implementation of the Plan is a successful practice of comprehensive research to break through major scientific problems, which has made outstanding achievements and improved people's understanding of continental evolution. The development trend of research in the field promoted by the project is as follows.

3.1 An Important Link in Global Continental Evolution: Craton Destruction

Through the comparative study of global cratons, it is found that lithospheric thinning is a common phenomenon in the evolution of global cratons, but craton destruction mainly occurs in the continental area adjacent to the subduction boundary of oceanic plates. The interaction between subducted plates, the lithospheric mantle and the asthenospheric mantle leads to craton destruction. The research results of the Plan draw the conclusion that craton destruction is an important link in global continental evolution, establish craton destruction theory, and promote the law summarized from the regional study of North China to the discussion of global problems.

As an important geodynamic process on the earth, craton destruction is of great significance to the formation and evolution of continents. In traditional geological research, why the continent was preserved was a problem of great concern to geologists at that time. The discovery of craton destruction made people begin to think about why the continent was destroyed. Eastern North China, western North America and western South America are currently recognized as typical areas of craton destruction, but this geodynamic process should have occurred many times in the history of the earth. The continent converged and cratonized after its initial formation and then tended to be stable, but this is not the end of continental evolution. Under the influence of the subduction of the surrounding oceanic plates, the craton will be destroyed. When the deep mantle returns to the normal state, the upper continent tends to stabilize, which completes a new process of continental stability. Therefore, craton destruction is an important link in continental evolution. The establishment of craton destruction theory has developed the plate tectonic theory and achieved a major theoretical breakthrough.

3.2 Global Research Hotspot: Craton Destruction

The NCC can be regarded as the cradle of geoscience in China. The "Yanshan Movement", "rejuvenation of platform" and "lithospheric thinning" all come from the study of the North China geology. The NCC destruction represents the most intense and typical craton destruction in the world. The Plan makes full use of China's regional advantages and long-term research accumulation. Based on excellent research results and innovative understanding, craton destruction has become a hotspot and frontier in the field of international earth research. At present, it is generally accepted that the NCC is a typical example of global craton destruction, and the NCC destruction is one of the important processes of global continental evolution. The Plan is a successful example of using China's regional advantages to explore the frontier of geoscience.

The NCC destruction was regarded as a "regional" scientific problem by Western scientists in the past. Through the implementation of the Plan, the scientific connotation of its continental evolution was revealed, and rapidly promoted it to be a global research hotspot. Chinese scientists have played an important role in promoting the research field of craton destruction, and some excellent scientists and research teams have stepped into the forefront of international solid geoscience research. The NCC destruction was a hotspot in the top 10 geoscience fields in the 2014 Research Frontier and 2015 Research Frontier published by Thomson Reuters-Chinese Academy of Sciences for two consecutive years. It is also the only geoscience research frontier dominated by Chinese scientists. The Plan has become a model for studying the evolution of global continents with regional examples (by using China's regional advantages), making the "regional" scientific issue of "the NCC destruction" a global research hotspot.

3.3 Birth of New Theories of Geoscience

The theory of plate tectonics based on the evolution of the oceanic lithosphere has encountered serious challenges in explaining continental geology, which urges people to understand the formation, evolution and dynamics of continents from a new perspective. According to the theory of plate tectonics, the intraplate is rigid and stable, and the tectonic deformation and magmatic activity are limited to the plate edges. However, the extensive and strong deformation and magmatism found in some continental plates cannot be explained by the traditional plate tectonic theory. With people's in-depth understanding of the composition, structure, rheology and deep processes of the continental and oceanic lithosphere, it is found that there are great differences between them, which urges people to re-examine and understand the law and dynamics of continental evolution. Therefore, the in-depth study of the formation, evolution and dynamics of continents has become the key entry point to give birth to a new theory of geoscience. It is of great scientific significance to take continental evolution as the breakthrough point and occupy the commanding height of the development of contemporary geoscience, because it will lead the major new development of earth sciences and related science and technology.

At present, geoscience is in a new period of major development and is facing a new opportunity for the innovation and development of geotectonic theory. All countries in the world, especially developed countries, are seizing commanding heights with the implementation of a series of national plans, competing and contributing to the establishment of a new major theory of geoscience. China should seize this opportunity, give full play to its unique advantages and strength conditions, and participate in the competition for the major development of earth sciences. At present, China has a series of favorable conditions to seize this opportunity, including a high-level research team and a first-class experimental platform in the continental evolution field; unique regional advantages and long-term accumulation of continental evolution research; geoscience research with international vision and global thinking and beginning to move toward the world; and social and economic development that can provide sufficient financial and material resources for high-level basic research and international cooperation. For this reason, China should gather its strength, focus on key core scientific issues, explore continental evolution and dynamics, build a new theoretical system of continental evolution, lead the development of geoscience and make original contributions to the new development of geoscience in the world.

The study of the NCC destruction is a successful example for joint research, which makes use of its regional advantages and aims at the forefront of its discipline development. This successful experience deserves further promotion. At present, the directions that should be focused on and are expected to make new breakthroughs include but are not limited to the following: the destruction and reworking of the continental margin in eastern China, reworking of the South China continent, continental rheology, the interaction between the continental margin and the oceanic plate, the transition between ocean and land and the fluid-rock interaction, the interaction of large mantle wedges, the deep carbon cycle and the dynamics of major events on the continental margins.

3.4 Multi-scale Physical and Chemical Observations

The processes of tectonic change, material exchange and interaction between the crust and the mantle in the continental lithosphere will inevitably lead to changes in lithospheric composition and structure, resulting in the redistribution of a series of elements and isotopes. The long and complex evolution process of the continental lithosphere is recorded in various minerals and rocks of different ages, especially inclusions derived from the lower crust and the lithospheric mantle, which contain direct information on the composition and structure of the lithosphere at some locations and are particularly valuable geological records. Due to the complexity of geological records and the general superposition characteristics, these inclusions must be carefully analyzed. The Plan explores the evolution law of the lithosphere of the NCC and reliably defines the evolution process, property transformation and evolution mechanism of the lithosphere through experimental study on mineralogy, petrology, geochemistry and geochronology of a large number of deep-seated inclusions (crust–mantle xenoliths) and intrusive rock samples, especially using the newly developed radioisotope (Hf, Os) and nontraditional stable isotope tracing technology.

The geophysical field, especially the seismic wave field, is the original data that propagate over the whole earth, fully carry the information inside the earth and can be received on the surface. Because many factors control the changes in physical fields, a high-precision, high-resolution and high-coverage observation system is needed to inversely infer the properties of matter and structure in the earth's interior from the wave field records obtained from the surface. The Plan has carried out comprehensive observations on the deep structure and tectonic characteristics of the NCC, covered the NCC and its adjacent areas on the plate, reached the upper mantle level at depth, and obtained a large amount of seismic data carrying the original information of the deep structure. Using the new observation data, the textural information of crustal and upper mantle seismic wave velocity distribution, velocity discontinuity structure and upper mantle anisotropy in the NCC were obtained from these high-precision, high-resolution and high-coverage seismic data. The detailed structure and state information of the crust–upper mantle provide an indispensable quantitative basis for understanding the scope, mode and dynamics of craton destruction.

The earth process and its interaction are complex, and its time scale ranges from earthquakes of a few seconds to the earth evolution of billions of years; its spatial scale spans more than ten orders of magnitude from the change in nanoscale mineral microdomain structure and composition to global plate tectonics on the scale of tens of thousands of kilometers. Information on the composition and evolution time can be obtained by high-precision isotopic dating, high-precision trace component testing, and microdomain and *in-situ* element and isotopic abundance analyses of rock samples. The characteristics of rock samples directly from the deep earth determine the important value of these results. However, the sampling points in a limited area greatly restrict the global judgment of tectonic properties. The tectonic information carried by the geophysical field has the advantage of covering the whole earth, but it lacks direct time and material records. Both geochemical analysis and geophysical observation are indispensable. Based on the geochemical analysis of the composition, structure and evolution age of the lithosphere on geological samples, the seismic exploration and structure imaging of the crust–upper mantle structure, and a detailed study of the geological evolution of tectonic belts, structures and outcrop geology, the Plan has clarified the properties of the lithosphere of the NCC and reveal the process of craton destruction against the background of four-dimensional spatiotemporal evolution.

The Plan has made a major breakthrough in the study of craton destruction, which has strongly confirmed the great improvement of detection ability brought by technological progress and promoted the continuous deepening of geoscience theoretical research. Careful observation and study of the texture, structure and material of the target area is the basis for the in-depth study of the scientific problems of continental tectonic evolution. The new exploration and testing capabilities developed on the basis of modern scientific and technological progress provide necessary research means for high-precision, high-resolution and high-coverage geophysical exploration and fine geochemical analysis. At the same time, high-quality observation data have been obtained by advanced data acquisition technology. The rapid development of computer technology has greatly improved the data processing and interpretation ability. With the development of earth sciences, scientists will pay more attention to the systematic collection, accumulation and analysis of basic scientific data by using modern observation, experimental testing and information technology to integrate cutting-edge research with the development of high and new technology.

3.5 Interdisciplinary Intersection and Integration

The study of continental evolution involves the in-depth exploration of complex geological processes. It is difficult for a single discipline and short-term research to make a major breakthrough in this field. Therefore, multidisciplinary collaborative research is the key to producing innovative achievements and making theoretical breakthroughs in the field of continental evolution.

The long-term and fruitful exchanges and cooperation in the implementation of the Plan not only were beneficial to solve scientific problems but also improved the multidisciplinary comprehensive research level of geology, geophysics, geochemistry, numerical simulation and high temperature and high pressure experiments in China, playing a demonstration and leading role for the intersection of different disciplines. In the implementation of the Plan, China's young generation of scientific researchers gradually changed from traditional geologists, geophysicists or geochemists to solid geoscientists, and their vision is more global and forward-looking. The working mode of interdisciplinary and large team cooperation has had a positive impact on changing the current scientific research situation of "breaking up into parts is easy, but gathering parts into unity is difficult". The implementation of the Plan has continuously formed a multidisciplinary comprehensive research team and has an important international impact.

Through the implementation of the Plan, the mode of organizing multidisciplinary and large teams to jointly tackle key problems has also had a positive impact in the field of geoscience. This mode is embodied as follows.

(1) Interdisciplinary intersection at the level of organization and management. The science department in charge of the Plan is the Department of Earth Sciences, the NSFC, and the relevant science departments are the Department of Mathematical and Physical Sciences and the Department of Information Sciences. Members of the management working group include members of the Department of Earth Sciences, the Department of Mathematical and Physical Sciences, the Department of Information Sciences and the Planning Bureau of the NSFC. The guiding expert group is composed of experts in geology, geophysics, geochemistry, and mathematical and physical sciences. Therefore, the Plan has promoted the

intersection and integration of different disciplines at the level of organization, management and guidance.

(2) Cross disciplines in top-level design and layout. The Plan pays great attention to substantive interdisciplinary intersection in terms of positioning, top-level design and project layout. The orientation of the Plan is to concentrate China's advantageous research forces in the fields of earth sciences, mathematical and physical sciences and information sciences, break through the constraints of traditional discipline boundaries, and carry out effective cross integration among different disciplines. Through the implementation of the Plan, the multidisciplinary comprehensive research level of geology, geophysics and geochemistry and the leapfrog development of geology, geophysics and geochemistry have been improved. In the top-level design, the overall scientific goal is realized through comprehensive and joint research of geophysics, geochemistry, geology, high temperature and high pressure experiments, numerical simulation and other disciplines. In terms of project layout, research projects of different disciplines are arranged. Therefore, the Plan is jointly undertaken by multidisciplinary scientists. The realization of the overall scientific goal and the solution to the core scientific problems depend on comprehensive and joint research among multiple disciplines.

(3) Interdisciplinary intersection in project implementation. The key scientific issues involved in the Plan often require multidisciplinary comprehensive research. Therefore, the members of the research groups within many projects are composed of researchers from different disciplines, and these projects have substantive interdisciplinary research in the implementation process. According to statistics, 72% of the project participants in the arranged projects are composed of multidisciplinary researchers.

(4) Interdisciplinary intersection in the exchange of achievements. During the implementation of the Plan, we also attach great importance to the communication and collision between different disciplines, and personnel from different disciplines discuss and exchange the same core scientific issues. At the annual year-end exchange meeting of major research plans, there is no division of subjects in the arrangement of speeches and exchanges. Project participants from different disciplines exchanged information and progress, held discussions with each other, and achieved good results. At national academic conferences in China, such as the annual "National Symposium on Petrology and Geodynamics" and "National Symposium on Structural Geology and Geodynamics", there are special topics on "North China Craton Destruction", which are also jointly attended and discussed by personnel from different disciplines. At relevant international academic conferences, such as "Craton Formation and Destruction (Beijing, 2011)" and "Mesozoic Continental Stretching Mechanism in East Asia (Orleans, France, 2010)", the discussion on the NCC destruction was also carried out jointly by personnel of different disciplines, which fully reflected the exchange, intersection and integration between different disciplines.

(5) Training of multidisciplinary talent and teams. The implementation of the Plan has promoted the cross integration of scientists from different disciplines,

different units, and even different countries and regions and gradually transformed China's young generation of scientific researchers from traditional geologists, geophysicists or geochemists to solid earth scientists. Their vision is more global and forward-looking. Driven by this research plan, a number of comprehensive research teams with great international influence in China's geoscience are forming, which are capable of tackling major and international cutting-edge scientific problems. At the same time, it has also driven the rapid development of China's geoscience research.

(6) Multidisciplinary natural laboratory. The NCC has become a natural laboratory for multidisciplinary comprehensive research in China and even in the world, attracting a large number of high-level scientists from different disciplines at home and abroad. These researchers from different disciplines carried out research on the common theme of the NCC destruction and regarded the NCC as a natural laboratory, which played a demonstration and leading role in the intersection and integration of disciplines.

Major Research Achievements

From the perspective of earth system sciences and based on the analysis technology and detection means of modern earth sciences, the Plan was carried out by Chinese scientists through systematic observations, experiments and theoretical analyses, with an emphasis on the spatio-temporal distribution range and process of the NCC destruction, the interaction and dynamic mechanism of different earth spheres during craton destruction, the resources and biological effects of craton destruction, and the cognitive level of the formation and evolution of continents. Through ten years of research and focusing on core scientific issues, we have successfully completed the predetermined scientific objectives and achieved the following innovative research results.

1 New Concept of "Craton Destruction"

The earth's continents consist of two basic geological units: orogenic belts and cratons. The most basic features of cratons are ancient formation ages and long-term stability. The craton is stable because of its extremely thick lithosphere with low density and low water content (Fig. 3.1). The cratonic lithosphere not only has strong rigidity but can also float on the asthenosphere, which can resist the reworking of various geological processes in the later stage to a great extent. Therefore, after the formation of most cratons in the world, there is basically no large-scale tectonic deformation, magmatism or seismicity, and there is a lack of large-scale endogenous mineralization. The NCC has an ancient rock record of more than 3.8 billion years, which has remained stable from 1.8 Ga to 200 Ma.

Since 200 Ma, especially in the late Mesozoic, the NCC has experienced frequent large-scale magmatism and strong crustal deformation, and its original craton structure and stability have been transformed and destroyed to varying degrees. The late Mesozoic tectonic evolution of the NCC shows that the craton can not only maintain long-term stability but also be destroyed. The geological phenomenon of craton

© Zhejiang University Press 2023

R. Zhu (ed.), *Research on Destruction of the North China Craton*, Reports of China's Basic Research, https://doi.org/10.1007/978-981-99-6046-0_3

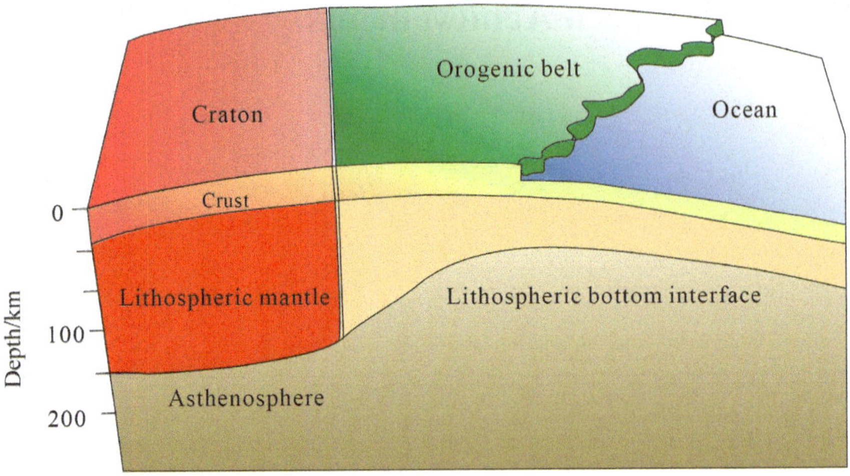

Fig. 3.1 Schematic diagram of craton lithospheric structure

destruction undoubtedly challenges the classical theory of plate tectonics. How the stable craton is destroyed has become a major scientific problem that needs to be deeply explored.

The large-scale crustal deformation and magmatism of the NCC in the Meso-Cenozoic have long been recognized by Chinese geologists. The concepts of the "Yanshan Movement" and "rejuvenation of platform" are put forward based on the study of Jurassic–Cretaceous tectono-magmatic activity in the NCC. With the deepening of research, "lithospheric unrooting", "lithospheric thinning" and "lithospheric delamination" have been put forward. In particular, the lithosphere in the eastern part of the NCC was thinned by more than 100 km in the late Mesozoic. Early scholars equated lithospheric thinning with craton destruction. The Plan reveals that the lithospheric mantle of the NCC has a nature of typical craton in the Paleozoic, generally up to 200 km thick, while the present thickness of its lithosphere in the east of the craton is only 60–80 km, which is similar to the oceanic lithospheric mantle in material composition and properties. Comprehensive studies of geology, geophysics, geochemistry and other disciplines show that large-scale magmatic activity, strong tectonic deformation and massive lithospheric thinning of the NCC are only manifestations in the process of craton evolution, and their essence is the fundamental transformation of material composition and physicochemical properties of the lithospheric mantle, that is, from a typical continental cratonic lithospheric mantle to a young oceanic lithospheric mantle. This great change in the properties of the lithospheric mantle leads to the loss of stability of the NCC. Therefore, the new concept of "lithospheric thinning is not equal to craton destruction" is clarified, and the definition of "Craton Destruction" is put forward. The craton destruction refers to the loss of its overall stability, due to the fundamental change in its lithospheric mantle attribute, and it is characterized by large-scale lithospheric thinning and strong magma-tectonic activities. There was no clear definition of craton destruction in the world before the Plan. The new concept of "Craton Destruction" proposed by the Chinese and recognized by international peers is of great pioneering significance. It clarifies the essence and scientific connotation of craton destruction, changes the traditional concept of "invariability" of ancient cratons, and improves people's understanding of continental evolution.

2 Spatio-temporal Distribution of Craton Destruction

2.1 Deep Structure of the Craton

Craton destruction is the product of continental evolution, and its controlling factor is deep in the earth. The structure and state of the crust–upper mantle is an important aspect to determine the destruction range of cratons, and it is also an indispensable basis to understand the mode and dynamics of craton destruction. Seismic exploration technologies have unique advantages in detecting deep structures of the earth. Using dense seismic arrays combined with active source seismic detection to detect the earth's internal structure, the researchers conducted a comprehensive exploration and research on the crust–upper mantle structure in North China and its adjacent areas. Based on the new high-resolution data obtained from seismic observations, the detailed crustal structure of representative sections was established, and the transformation of the crust caused by craton destruction was clarified. The distribution of lithospheric thickness in North China was ascertained, and the spatial scope of lithospheric thinning was determined. Direct evidence of the spatial pattern of craton destruction was obtained.

1. **Overall Exploration of Deep Structure and State**

In the Plan, dense transportable seismic arrays were fully deployed throughout the NCC and its periphery, and the structure and state of the crust–upper mantle were studied in combination with the active source seismic observation profiles (Fig. 3.2). Using advanced broadband seismic instruments, 8 profiles (station spacing of 10–15 km) and a two-dimensional array were successively deployed in the NCC and its adjacent areas, with a total of 688 stations of transportable seismic arrays. Three wide-angle reflection/refraction profiles with long-range artificial seismic source (3650 km) and two seismic observation profiles using the combined ocean-bottom-seismometer (OBS) and land portable seismometer (860 km) were completed. These deep explorations cover the NCC and its adjacent areas, and reach the upper mantle at depth. The seismic record data carrying the original information of deep structures were obtained. In addition to these new observation data, the natural seismic records of approximately 370 stations in China National Digital Seismic Networks and 43 artificial seismic wide-angle reflection/refraction profiles in the previous stage were collected. Through the development and application of effective seismic imaging technology, the structure information such as the crustal and upper mantle seismic

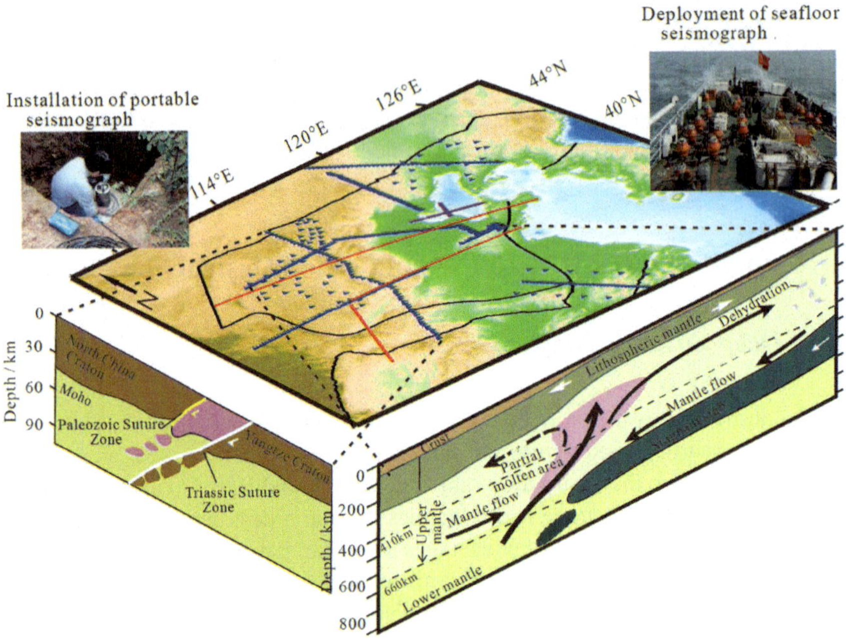

Fig. 3.2 Active source seismic observation profile and station distribution of transportable arrays in North China. Blue triangle: transportable seismic station. Red line: wide angle reflection/refraction deep seismic sounding line. Purple stripe: submarine seismic observation station

wave velocity distribution, the structure of velocity discontinuity interfaces, and the upper mantle anisotropy in North China were obtained from seismic data. Finally, the database and application system of three-dimensional velocity structure "Crustal and Upper Mantle Velocity Model of North China" are established. The structure model includes the data and images of buried depth and stratification velocity at each velocity interface of the crust, the lithosphere and the upper mantle, the data and images of upper mantle velocity disturbance, and the crustal velocity structure of typical profiles. The structural model and database have been on the website (http://www.craton.cn/data) for sharing.

2. **Variation Law of Lithospheric Thickness of Craton**

Lithospheric thickness is a basic attribute representing the evolution state of a craton. Based on the original teleseismic waveform data recorded at stations within the temporary array and fixed network, the thickness distribution of the lithosphere of the NCC has been obtained by using a wave equation based receiver function migration method (Fig. 3.3). Combined with the results of geological and geochemical studies, it is found that there are significant differences in the structure and properties of the crust and the lithospheric mantle between the east and west of the NCC. The crust (thickness <35 km) and the lithosphere (60–80 km thick) of the eastern block are obviously different from the typical cratonic lithospheric structure. The

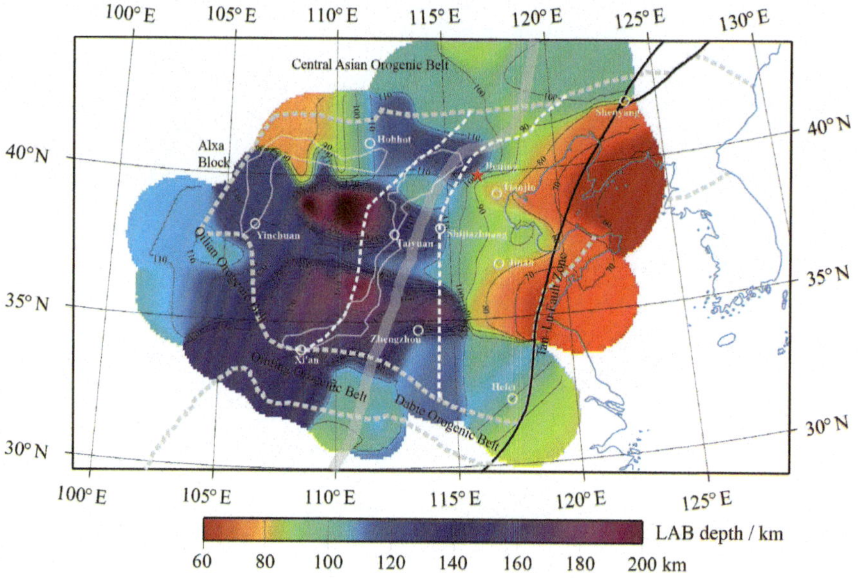

Fig. 3.3 Lithospheric thickness (LAB depth) distribution of the North China Craton. The value on the isoline represents the lithospheric thickness (km); the dotted line indicates the boundary of tectonic divisions; and the gray zone is a north–south gravity gradient zone

crust (approximately 45 km thick) and the lithosphere (up to 200 km thick) of the western block maintain the characteristics of a typical craton, while the thickness and structure of the crust and the lithosphere of the central continental block show strong lateral heterogeneity. The lithospheric bottom interface beneath the eastern part of the NCC is a strong velocity discontinuity surface, and there is an internal lithospheric discontinuity surface with seismic wave velocity decreasing with depth at 80–100 km deep in the west. The thickness and properties of the lithosphere under the NCC are highly laterally heterogeneous. The lithosphere of the eastern part of the NCC is "thin" as a whole, and in the central to western part is mainly "thick" with "thin" locally, which indicates that the lithosphere in the eastern part of the NCC has been significantly thinned as a whole, while the lithosphere in the central to western part has only been thinned locally. Therefore, it is clear that lithospheric thinning of the NCC mainly occurred to the east of the Taihang Mountains.

3. **Crustal Transformation Caused by Craton Destruction**

The crust at the outermost layer of the earth is under relatively low temperature and pressure, which can retain the record of long-term tectonic evolution to the greatest extent. Using the newly obtained seismic observation data of transportable arrays and fixed stations and collecting the seismic detection results of multistage active sources, we determined the basic characteristics of the crustal structure of the NCC (Fig. 3.4). The present crustal structure of the NCC shows a general pattern of east–westward change, and roughly is bounded by the north–south gravity gradient

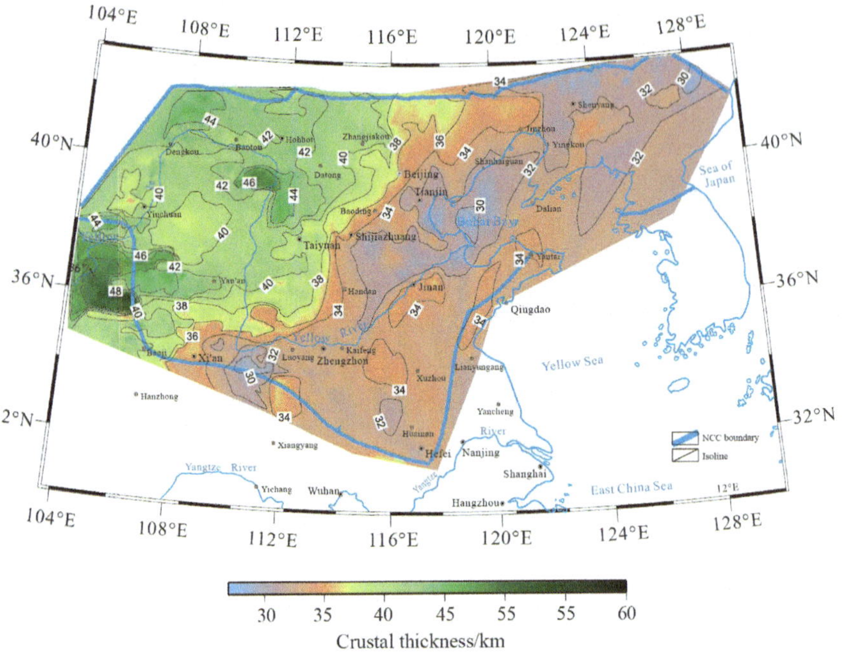

Fig. 3.4 Crustal thickness distribution of the North China Craton. The values of color columns and isolines in the figure represent the crustal thickness (km); the blue line indicates the NCC boundary

zone. The eastern NCC is characterized by a thin crust, a thick sedimentary cover, a relatively smooth Moho, and horizontally extended inter-layering of high and low velocities in the crust, which represents that the extension and thinning of the crust are transformed by large-scale ductile deformation structures. From east to west, the crust gradually thickened and changed into a flat layered structure, which has the characteristics of typical cratonic crust. On the northern side of the central NCC, the inclined low-velocity zones and the crustal root represent the tectonic remnant of Paleoproterozoic continental collision and amalgamation. On the southwestern margin of the NCC, the crust significantly thickens due to the eastward expansion of the Tibetan Plateau.

The detailed crustal structure profile is constructed by dense transportable seismic array detection. As an example, Fig. 3.5 shows the S-wave velocity structure of a profile crossing the northern part of the NCC along the east–west direction. The detailed crustal structure of several profiles shows that the lower crust of eastern North China is characterized by extensive low-velocity zones, indicating that crustal remelting and weakening occurred. At the same time, there is no image of quasi-symmetrical distribution or centralized distribution in the low-velocity zone in the crust, and there is no structural record of large-scale mantle plume action. The seismic profile has proven the middle and lower crustal structure of the metamorphic core complex area in the Liaodong Peninsula, the sedimentary caprock structure of the

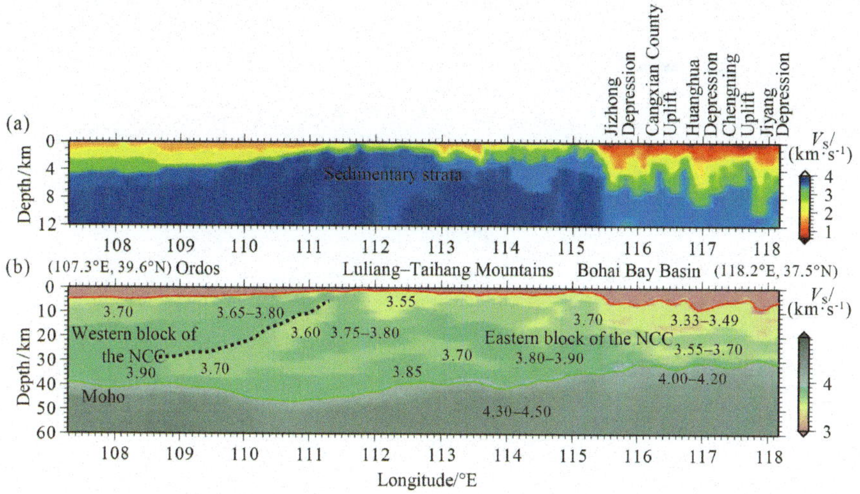

Fig. 3.5 S-wave velocity structure of Lijin, Hebei and Otog, Inner Mongolia. **a** The stratification structure of sedimentary strata is shown in detail in the part with a depth of more than 12 km; **b** the value of S-wave velocity (V_S, km·s^{-1}) is marked by the color column and number

evolution of the Bohai Bay Basin from fault depression to depression basin, and the Moho dislocation under the Tan–Lu Fault Zone. These structural features indicate that strong magmatism and extension transformed the crust in the eastern part of the NCC.

2.2 *Spatio-temporal Distribution of the North China Craton Destruction*

The fundamental change in material composition and physicochemical properties of the lithospheric mantle will lead to the destruction of the inherent stability of the craton, which is the essence of craton destruction. Large-scale magmatism, strong tectonic deformation and massive lithospheric thinning are manifestations of craton destruction. These manifestations are the basis for determining the spatial range of craton destruction. The systematic shallow geological study shows that the shallow crust of the eastern NCC was characterized by a series of extensional faults, extensional basins and metamorphic core complexes during the Early Cretaceous, showing extensive crustal extensional deformation. The western part of the NCC shows a large basin pattern of overall depression or fault-depression combination against the background of weak extension, and the weak magmatic and fault activities reflect the stable cratonic state. From the late Paleozoic to Triassic, magmatic activities of the NCC occurred in the northern margin. The magmatic activity in the Jurassic was mainly concentrated in the northern and southeastern margins. The range and intensity of magmatism reached their peak during the Early Cretaceous and occurred

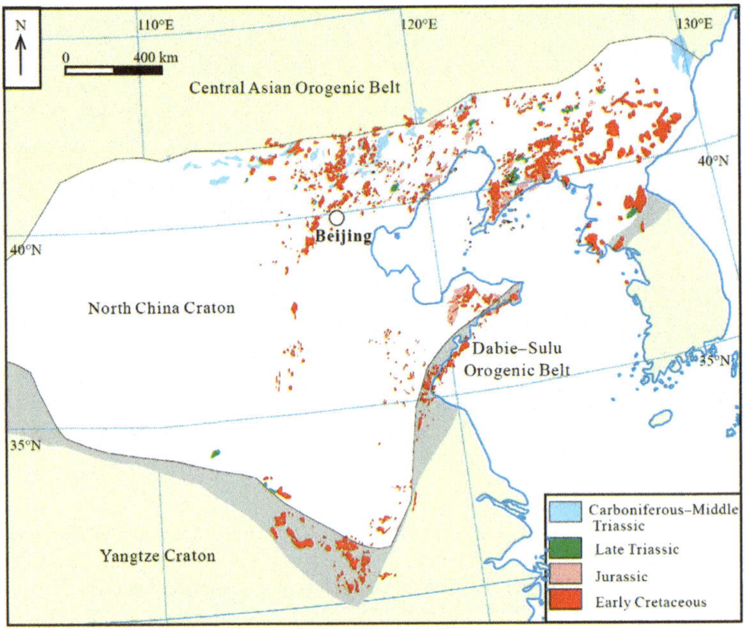

Fig. 3.6 Distribution of Mesozoic magmatic rocks in the North China Craton

throughout the eastern craton (Yanshan Tectonic Belt, Taihang Mountains, Dabie–Sulu Orogenic Belt, North China Platform and both sides of the Tan–Lu Fault Zone) (Fig. 3.6). Large-scale magmatism occurred in the eastern part of the NCC in the Mesozoic, and the resultant magmatic rocks were distributed in a dispersed manner, indicating that the NCC lost its original overall stability; that is, craton destruction occurred in the NCC.

Through the seismological exploration of the deep structure in North China, the thickness change of the lithosphere of the NCC was obtained. There are significant differences in the structure and properties of the crust and the lithospheric mantle in the east and west of the NCC. The crust (thickness <35 km) and the lithosphere (60–80 km thick) of the eastern block are obviously different from the typical cratonic lithosphere structure, and the crust (approximately 45 km thick) and the lithosphere (up to 200 km thick) of the western block maintain the characteristics of the typical craton, while the thickness and structure of the crust and the lithosphere of the central continental block show strong lateral heterogeneity. The change of lithospheric thickness provides direct evidence of the spatial range of lithospheric thinning. The above observations and studies confirm that the eastern part of the NCC has experienced overall destruction and loss of stability, while the western block still maintains the attribute of overall stability, and the central part has not completely lost the attribute of overall stability (partial transformation and reworking of its lithospheric structure and properties).

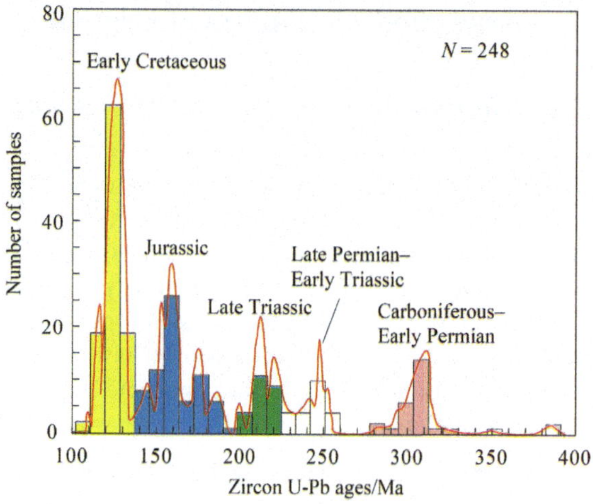

Fig. 3.7 Distribution of phanerozoic magmatism in the North China Craton

After chronological measurement and geochemical analysis of a large number of rock samples, the six stages of magmatism of the NCC in the Phanerozoic can be identified based on the systematic study of the spatio-temporal distribution, origin and tectonic background of igneous rocks (Fig. 3.7). Among them, alkaline basalts in the Cenozoic only occasionally appear. The first four stages of magmatism occurred only along the marginal areas in response to the amalgamation between the NCC and the blocks to the north and south sides. The middle Early Cretaceous is the peak period of magmatism in eastern North China, which is characterized by large-scale high-K calc-alkaline magmatic activities and comes from the enriched lithospheric mantle. The systematic study of the tectonic evolution of the NCC shows that the compressional regime occurred before the destruction. The Early Cretaceous destruction period corresponds to the strong extension of the shallow crust. A compression event at the end of the Early Cretaceous ended the peak destruction in the eastern NCC.

Through the study of the transformation process of lithospheric physico-chemical properties and a large number of studies of high-precision chronologies in the NCC, combined with the signs of magmatism, crustal deformation and extensional basins, the researchers determined that the peak period of the NCC destruction was the Early Cretaceous (the peak was *ca.* 125 Ma). After the destruction peak period, the eastern part of the NCC completely lost the stability of the typical craton.

In short, multidisciplinary comprehensive analysis and research reliably showed that the destruction of the NCC occurred to the east of Taihang Mountains, and the peak period of the destruction was the Early Cretaceous.

3 Process and Mechanism of Cratonic Destruction

3.1 Shallow Tectonic Process of Cratonic Destruction

The systematic and in-depth study of shallow geology can make an important contribution to the understanding of the spatio-temporal distribution, destruction process, shallow effect and dynamic mechanism of the NCC destruction. Therefore, since the implementation of the Plan, structural geology, basin and oil and gas geology, sedimentology, stratigraphy and paleontology have been used to deeply study deformation characteristics, the development law of metamorphic core complexes, the tectonic evolution history, the formation and evolution of basin in the NCC since the Mesozoic, and many important advances have been made.

1. **Tectonic Pattern Before the Peak Destruction of the Craton**

Through the systematic study of the tectonic evolution of the NCC, it is found that the craton experienced multistage compressive deformation during the Mesozoic before the peak destruction of the Early Cretaceous. During that time, terrestrial depression basins were developed, but the overall stability was basically not lost. The final closure of the Paleo-Asian Ocean (Solonker Ocean) on the northern side of the NCC from the end of the Paleozoic to the Middle Triassic led to the strong shortening deformation (hinterland deformation) of the Yinshan–Yanshan Tectonic Belt in the north of the NCC. The tectonic belt was still in an active continental margin environment in the late Paleozoic, and a series of arc magmatism developed due to the southward subduction of the Paleo–Asian Ocean. During the collision of the South China plate and the NCC along the Qinling–Dabie–Sulu orogens in the Middle Triassic, the southern margin and eastern side of the NCC were also involved in hinterland deformation, but there was no arc magmatism before the collision. At the end of the Middle Jurassic and the beginning of the Early Cretaceous, two compression events occurred again on the NCC, called Episodes A and B of the Yanshan Movement, respectively, which belong to the active marginal deformation caused by the low-angle subduction of the Paleo-Pacific Plate (Izanagi Plate) (Fig. 3.8). During the Yanshan Movement, the Tan–Lu Fault Zone was strongly active, and a series of fault structures were associated in the east of the NCC, forming a series of fault blocks. Between the above four compression events, large depression or flexure-type terrestrial basins (T_1–J_2) developed in the NCC. At present, the basins are well preserved in the Ordos Basin, and those in the eastern NCC are locally residual due to late uplift and strong deformation. The overall subsidence characteristic of the craton indicated by these basins reflects that the overall stability of craton was not destroyed at that time.

2. **Shallow Structural Features in Peak Destruction**

A large number of studies showed that the peak destruction and lithospheric thinning of the eastern NCC took place in the Early Cretaceous, corresponding to the strong extension of the shallow crust. With a large number of volcanic eruptions and

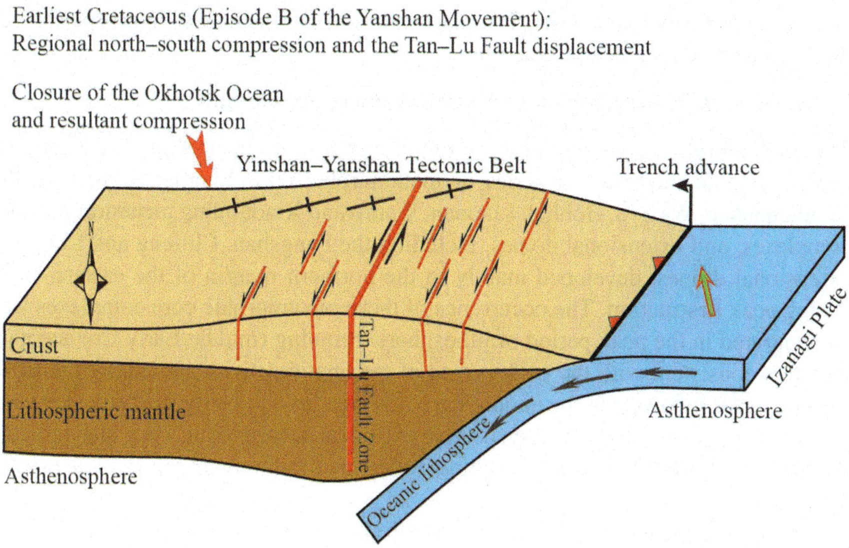

Fig. 3.8 Dynamic model of the early Cretaceous Episode B of the Yanshan Movement in the eastern NCC

magmatic emplacement, the shallow crust has a high heat flow value, which is characterized by a series of extensional faults, extensional basins and metamorphic core complexes. During this period, the metamorphic core complex in the northern NCC appeared alternately with faulted basins, showing a basin-range tectonic pattern. A series of Early Cretaceous extensional basins, such as Zhoukou, Guzhen, Xinyang–Huangchuan, Hefei and Jiaolai basins, developed in the southern margin of the eastern NCC. The interior of the eastern NCC is mainly composed of small faulted basin groups, including southwestern Shandong basin groups (the Qufu, Sishui, Pingyi, Dawenkou, Xintai, Mengyin, Laiwu basins, etc.) and the Bohai Bay Basin groups. The Tan–Lu Fault Zone also changed into a giant extensional fault in this period, and a series of graben and half-graben faulted basins developed in its interior and margins. These continental extensional basins in the eastern NCC are filled with both a large number of clastic rocks and large-scale intermediate-to-acid volcanic rocks. The normal faults controlling development of the basins mainly strike north–north–east (NNE).

During the Early Cretaceous, the Ordos Basin in the western NCC was a large basin extending in a north–south direction. The western boundary of the basin is a large normal fault dipping toward the east, and the eastern boundary of the basin is mainly an overlap boundary. The sediments are thick in the west of the basin and thins to the east, showing a large-scale half-graben pattern with faulting in the west and overlapping in the east. Hetao Basin extending from east to west developed in the north of the basin, while there are still the small-scale Yinchuan and Bayanhaote basins in the west of the basin. In general, the western NCC presented a large basin pattern of overall depression or fault-depression combination in the weak extensional

Various geological phenomena indicate that the eastern NCC was not in the Andean/Cordillera-type continental margin arc or magmatic arc environment (also known as subduction-type orogenic belt environment) in the late Mesozoic, but in the back-arc environment near the continental side. The intense magmatism under the eastern NCC in the Early Cretaceous should not be the typical arc magmatic activity. It occurred in the lithosphere extension against the background of subduction plate retreat in the back-arc area of active continental margin.

3.2 Lithospheric Evolution of the North China Craton

The essence of cratonic destruction is the loss of its stability, which is due to the transformation of cratonic lithosphere into oceanic lithosphere. In the implementation of the Plan, the magmatic products of all periods of the Phanerozoic and its xenoliths have been studied in mineralogy, petrology and geochemistry. The study area spans major units, such as the eastern block of the NCC, the western block, the central zone and the Tan–Lu Fault Zone. The results of this series of studies have ascertained the attributes of the Paleozoic, Mesozoic, and Cenozoic lithosphere in North China. In addition, detailed petrological and in-situ microdomain analyses of the mineral zoning of peridotite xenoliths and xenocrysts with typical reaction structures were carried out, and the interaction process between mantle peridotite and melt as well as the composition and source of melt were discussed. These results reveal the Phanerozoic evolution of the NCC lithosphere and lay a foundation for understanding the nature of cratonic destruction.

1. **Spatial–temporal Difference of Meso-Cenozoic Magmatism and Crustal Recycling**

There is a close genetic relationship between magmatism and lithospheric evolution. Systematic mineralogical, petrological, geochronological and geochemical studies of Meso-Cenozoic magmatic rocks in the eastern NCC show that they have high chemical heterogeneity, which is related to various forms of recycling of crust-derived components. The magmatism and its genesis before and after 100 Ma are significantly different and have obvious stages.

Before 100 Ma, the NCC was characterized by large-scale high-potassium calc-alkaline magmatic activity, and the middle Early Cretaceous was the peak period of its activity. The magma of this period derived from the enriched lithospheric mantle is depleted in high-field strength elements and is composed of enriched Sr–Nd isotopes. The enrichment-metasomatism process is related to crustal recycling in various forms, such as deep subduction of the continent and delamination of the lower crust. It corresponds to the history of peripheral plate activity. The transformation of the lithospheric mantle in the NCC Lithosphere by the deep subduction crust material and its releasing fluid from the Yangtze continent determines that the lithospheric mantle in the core of the NCC has the characteristics of EM1-type, while the lithospheric mantle around the NCC has the characteristics of EM2-type. Mantle xenoliths from the Cenozoic Hannuoba basalts in Hebei Province record the lithospheric mantle melt-peridotite reaction induced by the subduction of the Paleo-Asian Ocean at 315 Ma and the melt-peridotite reaction caused by the partial melting of the recycled continental crust during 170–80 Ma. The geochemical characteristics of the Early Cretaceous alkaline picrites and high-magnesium basalts of Feixian, Shandong and Sihetun, western Liaoning, and the anti-ring zoning of clinopyroxene xenoliths indicate that the source area does not contain olivine or has less olivine. However, it contains many eclogite components, which indicates a significant contribution from crustal sources. The existence of a large number of crust-derived components in the late Mesozoic deep lithosphere of North China contributed to the occurrence of later large-scale melting events and the lithospheric thinning.

Since 100 Ma, intraplate basic magma has appeared in the NCC. The magma is enriched in high-field strength elements and is composed of relatively depleted Sr–Nd isotopes from the asthenospheric mantle. There are a large number of eclogite (and garnet pyroxenite) components in these basic magma source areas, which were mainly derived from the subducted oceanic crust metamorphosed into eclogite or garnet pyroxenite. A melt-peridotite reaction occurred between the adakitic to felsic melt produced by the partial melting of the subducting oceanic slab and the overlying MORB-type mantle wedge peridotite. Based on the analysis of deep geophysical data, it is inferred that these subducted oceanic crust components originated from oceanic slabs trapped in the mantle transition zone. The basalt composition younger than 20 Ma shows that the North China and Southeast China coasts are EM1- and EM2-type mantles, while the inland is the EM1-type mantle. Since the formation of the EM2-type mantle is related to the metasomatism of subducting fluids, the zoning pattern of the mantle also shows the influence of Pacific subduction on the evolution of the lithosphere.

Therefore, the characteristics of Meso-Cenozoic magmatism in the NCC have recorded the transformation effects of various crustal recycling (such as deep continental subduction, lower crustal delamination, and oceanic crust subduction) on the deep mantle of the NCC. It truly reflects the multistage evolutionary history of the interaction between the NCC and the surrounding plates and is also of great significance for revealing the evolution mechanism of the NCC lithosphere.

2. **Typical Craton-Type Lithospheric Properties Before Thinning**

The NCC is one of the oldest cratons in the world. It has an ancient continental crust of 3.8 Ga. At the end of the Paleoproterozoic (1.85 Ga), the eastern and western nuclei of the Archean collided to form a unified craton, leading to complete cratonization. Since then, the NCC has been in a stable tectonic state, which lasted until the early Paleozoic. The zircon geochronology and Hf isotope geochemistry of granulite xenoliths reveal that the lower crust of the NCC was mainly formed in the Archean. The eruption of Middle Ordovician kimberlites was the first magmatic activity after the NCC cratonization. Since the lithospheric mantle is mainly composed of peridotite with olivine as the main component mineral (>60%), the composition of olivine in peridotite from the NCC lithosphere can provide direct information on the composition of the lithospheric mantle to some extent. The typical craton-type lithospheric mantle (Archean lithospheric mantle) is characterized by highly refractory harzburgite and refractory lherzolite. In general, from the Archean to the Proterozoic and then to the Phanerozoic, the major element composition of lithospheric mantle peridotite will change from highly refractory to less refractory or even full, and the lithospheric mantle will change from harzburgite-dominated to lherzolite-dominated. The Archean continental lithospheric mantle is considered to be the remnant of highly partial melting of the original mantle at a depth of greater than 150 km. Therefore, there are obvious differences in mineral composition between Archeozoic and Phanerozoic lithospheric mantle. Compared with the chemical composition of the primitive mantle or asthenosphere, the Archean continental lithospheric mantle has basaltic components with highly depleted Fe, Ca and Al, which makes the lithospheric mantle highly refractory, and its density is lower than that of the underlying asthenosphere. Therefore, the Archean continental lithosphere can "float" over the asthenosphere for a long time, which is the main reason for the long-term stability of the Archaean cratonic lithosphere.

The age of the lithospheric mantle is difficult to determine in solid earth sciences. Traditionally, the age of the lithospheric mantle is estimated based on the depletion degree of the mantle. The F_0 value of olivine in mantle peridotite is taken as a sign. The higher the F_0 value is (the more refractory it is), the older the formation age may be. For example, the F_0 value of Archean mantle peridotite (>92) is higher than that of Phanerozoic mantle peridotite (<91). This is consistent with the fact that in the early stage of the earth's formation, the geothermal gradient was high, and the degree of partial melting was also high. This result suggests that the F_0 value of olivine in mantle peridotite represents not only compositional information but also age information to some extent. The more refractory the composition is, the older the formation age may be. Olivine xenocrysts, peridotite xenoliths and olivine inclusions in diamond-bearing kimberlites from Mengyin, Shandong and Fuxian, Liaoning (Fig. 3.9) all have high F_0 values (94–92), which are completely distributed in a range of the global Archean tectonic domain, suggesting that the lithospheric mantle represented by these olivines was probably formed in the Archean.

Re-Os isotopes are the most effective method for dating peridotite in the lithospheric mantle. Theoretically, the higher the degree of depletion of the lithospheric mantle is, the lower the Os isotope ratio ($^{187}Os/^{188}Os$) and the older the isotope age may be. The Os model age and Re depletion age of the peridotite enclaves from diamond-bearing kimberlites in Mengyin, Shandong and Fuxian, Liaoning also indicate that the NCC lithospheric mantle was formed in the Archean and preserved at least until the eruption of the Ordovician kimberlite.

Before the destruction of the NCC, it has typical craton-type continental lithosphere characteristics:

(i) The lithosphere is thick, approximately 200 km. (ii) The ancient lithosphere was formed in the Archean, older than 2.5 Ga. Both spinel and garnet facies in the lithospheric mantle were formed in the Archean. (iii) The lithospheric geothermal gradient is low, equivalent to 40 mW/m^2. (iv) The lithospheric mantle is refractory. A strong depletion event occurred in the Archean, forming magnesium-rich peridotite,

Fig. 3.9 Diamond-bearing kimberlite in Mengyin, Shandong, China. Kimberlite is the parent rock of diamond. The crystal, as a clear mineral particle in the picture, is diamond, which is produced in kimberlite in Mengyin, Shandong, China. Diamond is a natural mineral and the raw stone of diamond. Natural diamond is a simple crystal composed of carbon. It is formed in the conditions of high pressure and high temperature in the deep earth (approximately 180 km underground)

which is mainly composed of refractory harzburgite. The F_0 value of olivine is 94–92. (v) Because the lithospheric mantle is rich in magnesium and poor in iron, it has low density, high buoyancy and high seismic wave velocity, making it float on the asthenosphere for billions of years without being damaged. (vi) There are 80–100 km metasomatic enrichment layers in the lithospheric mantle. As the small-scale melt adiabatically rises to 80–100 km, it encounters a large thermal resistance barrier. The melt cannot continue to rise, replacing the lithospheric mantle to form an enrichment layer. (vii) Sr-Nd-Hf isotopes in the lithospheric mantle are weakly enriched. (viii) There is no large-scale magmatic activity, strong tectonic deformation and large earthquakes, and only a small amount of kimberlite eruption (the kimberlites in Mengyin, Shandong and Fuxian, Liaoning contain diamond and peridotite xenoliths).

The temperature and pressure calculation results from the Paleozoic kimberlite and garnet peridotite xenoliths, mantle mineral xenocrysts, heavy minerals and diamond solid inclusion mineral assemblages in the NCC and the Re-Os isotopic data of peridotite xenoliths all show that there was an ancient lithosphere with a low geothermal gradient and thickness of 200 km under the eastern NCC during the Paleozoic. At that time, the lithospheric mantle was mainly composed of refractory harzburgite and lherzolite, with depletion of major elements and slight enrichment in Sr–Nd isotopic composition, which had the characteristics of the typical craton-type lithospheric mantle.

3. **Transformation of Lithospheric Properties During Mesozoic Destruction**

The Paleozoic lithosphere of the NCC still has typical characteristics of a craton-type continental lithosphere. The destruction of the NCC mainly occurred in the Mesozoic. It experienced tectonic transformation and strong craton destruction in the Mesozoic. Therefore, the composition and tectonic state of the Mesozoic lithosphere are the key to understanding the destruction of the NCC and the thinning of the lithosphere. Because the Mesozoic mantle-derived magmatism mainly comes from the lithospheric mantle which carries fewer mantle xenoliths and xenocrysts, the products have become the key samples to explore the composition of the lithospheric mantle under the NCC in the Mesozoic. The Mesozoic lithospheric mantle has high chemical heterogeneity and obvious spatio-temporal variations.

Through the study of xenoliths from Mesozoic basalts in the NCC, we have proposed that the peridotite-melt reaction is an important mechanism for the transformation of the lithospheric mantle composition. The zonal peridotite xenoliths and pyroxene xenoliths carried by the Early Cretaceous basalts in Fangcheng County indicates that the two main minerals in the peridotite can react with the melt and cause a decrease in the mineral $Mg^{\#}$. The discovery of zonal peridotite xenoliths from the Late Cretaceous basalt in Jiaozhou further proves that the peridotite itself can react with the melt and form low-$Mg^{\#}$ peridotite or pyroxene peridotite. Obviously, these mineral xenoliths or peridotite xenoliths are the result of the reaction with the host basalt during the ascent process, thereby preserving its zonal structure. However, in the high-temperature environment of the lithospheric mantle, due to the high diffusion rate of Fe–Mg, this reaction will quickly achieve chemical equilibrium between the minerals. Therefore, it is difficult to preserve direct mineral reaction structures in mantle peridotite xenoliths.

Global comparison shows that the NCC has been the most severely destroyed area in the world since the Phanerozoic, which is a typical example of global craton destruction. The lower crust and lithospheric mantle structure in the western part of the NCC (west of the Taihang Mountain gravity lineament) roughly maintained the state before craton destruction and preserved a lithosphere approximately 200 km thick. The central and eastern parts of the NCC (to the east of the Taihang Mountain gravity lineament) experienced different degrees of craton destruction, lithospheric thinning and property transformation.

The thickness and composition of the Mesozoic lithosphere in the eastern part of the NCC have undergone significant changes. The lithospheric mantle has been modified to varying degrees from recycled crustal materials. The lithospheric mantle in the eastern NCC has undergone strong melting/fluid transformation and has changed from a refractory craton-type mantle to a fusible enriched mantle, with only a small amount of ancient mantle remnant. In contrast, the melting degree in the central part of the NCC is relatively weak, and the ancient mantle is abundant. The ancient lithospheric mantle in the western NCC has not undergone significant melt reworking.

Compared with the Paleozoic, the thickness of the lithosphere in the eastern NCC was reduced by 80–120 km during the Mesozoic (the crust and the lithospheric mantle were both significantly thinned). The lithospheric mantle was mainly composed of peridotite and pyroxene, presenting the characteristics of a relatively full composition of main elements and a relatively enriched Sr–Nd isotope composition. The lithospheric mantle has the remarkable transformation of material composition in the NCC during the Mesozoic. It also experienced the modification by recycled crustal materials, which produced obvious spatio-temporal heterogeneity. Zircon geochronology and Hf isotopes of the lower crustal granulite xenoliths reveal that the ancient lower crust of the NCC has also been subjected to multiphase and multistage magmatic underplating since the Phanerozoic, including the Caledonian and Hercynian, Indosinian, Yanshanian and Himalayan periods, and experienced a process of compositional transformation similar to that of the lithospheric mantle (Fig. 3.10). The structural characteristics of the Cenozoic lithosphere in the Hannuoba area, Hebei, are inferred from the petrological research results of mantle peridotite, pyroxenite and granulite xenoliths and the calculation results of mineral temperature and pressure. The rock types from the top of the lithosphere to the upper part of the lower crust are spinel lherzolite, garnet lherzolite, garnet pyroxenite, black pyroxenite, banded granulite and normal ancient granulite. Among them, spinel lherzolite is the rock at the top of the normal lithosphere, garnet lherzolite is the product of the reaction between melt and peridotite, various garnet pyroxenite and black pyroxenite are the products of magmatic underplating, and banded granulite is the product of the reaction and recrystallization of underplating magma and ancient granulite. Magmatic underplating mainly occurred at the bottom of the lower crust, resulting in the old upper and new lower structures of the lower crust. In short, the destruction of the NCC is manifested not only in the large-scale thinning of the lithosphere thickness but also in the great changes in the composition, nature and structure of the lithosphere and extensive magmatism. The destruction degree of the NCC reached its peak in the late Mesozoic, and then the eastern craton no longer has stable cratonic properties.

4. **Cenozoic Oceanic Lithospheric Mantle**

A large number of basaltic rocks have been exposed in the NCC since the Late Cretaceous. Among them, there are many outcrops of Cenozoic basalts, which are widely distributed, although a single outcrop area is small. Cenozoic basalts are mainly distributed along the Tan–Lu Fault Zone, the northern margin of the NCC and the Taihang Mountains. These basalts are mainly alkaline basalts and tholeiites, in which mantle-derived peridotite and pyroxenite xenoliths are generally contained in the alkaline basalts. These mantle xenoliths provide good samples for us to explore the compositional characteristics and evolution of the lithospheric mantle since the Late Cretaceous in the NCC.

Represented by mantle peridotite xenoliths in Qingdao, Shandong, low-$Mg^{\#}$ peridotite is similar in structure, petrology and mineralogy to Cenozoic low-$Mg^{\#}$ peridotite in other parts of the NCC, which represents the juvenile lithospheric mantle. The petrological and mineralogical characteristics of the high-$Mg^{\#}$ peridotite are

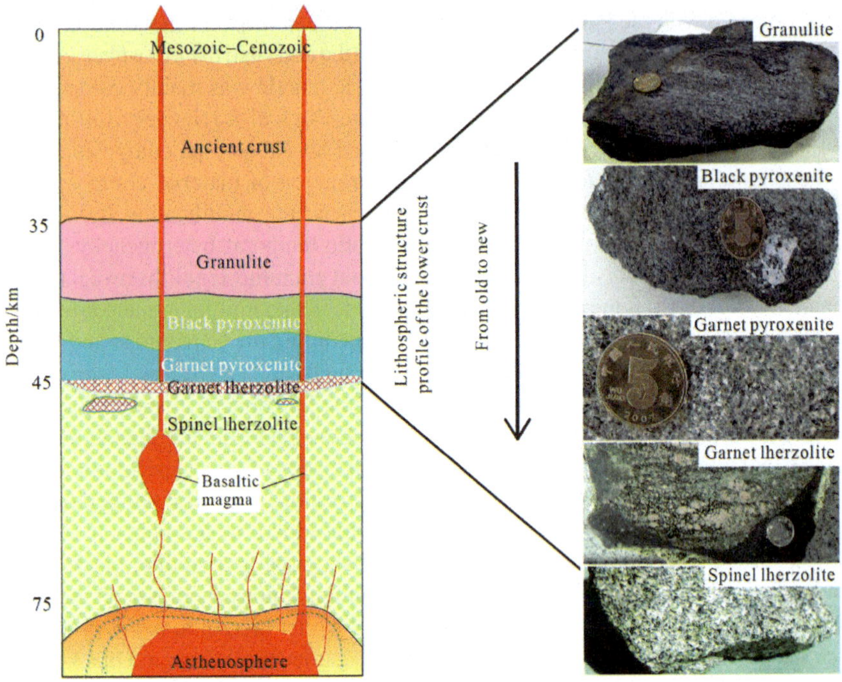

Fig. 3.10 Cenozoic lithospheric structure profile in the Hannuoba area within the North China Craton

similar to those of the ancient lithospheric mantle, so it represents the remnants of the ancient lithospheric mantle. Moreover, the high-Mg$^{\#}$ peridotite has also undergone complex mantle metasomatism or interactions between melt and peridotite.

The Beiyan area located in the Tan–Lu Fault Zone in the eastern part of the NCC is quite special. The mantle peridotite xenoliths obtained in this area can be divided into three types: lherzolite, clinopyroxene-rich lherzolite, and wehrlite. The F_0 value of peridotite is low (91.0–88.8), and its petrological and geochemical characteristics are similar to those of the late Mesozoic low-Mg$^{\#}$ peridotites in Qingdao and Junan and most of the Cenozoic basalts in the eastern NCC, which shows that the Beiyan lherzolite represents the new lithospheric mantle. The F_0 values (87.6–81.0) of clinopyroxene-rich lherzolite and wehrlite are lower than that of lherzolite, suggesting that these two low-F_0 peridotites were derived from reactions between lherzolite and melt. This indicates that the juvenile lithospheric mantle in this area continued to be modified by the interaction between peridotite and melt. Because the Beiyan basalt is located in the Tan–Lu Fault Zone, the upwelling of the asthenosphere in the Late Cretaceous to Eogene and the shearing along the fault zone provide the source and rising channel for the melt-peridotite reaction. The transformation of the lithospheric mantle in the fault zone is more intense, resulting in the main body of the lithospheric mantle along the Tan–Lu Fault Zone being dominated by full peridotite, which has the characteristics of the oceanic lithospheric mantle.

The study of the lherzolite xenoliths carried by Cenozoic basalts shows that the lithosphere in the eastern part of the NCC has thin thickness (<80 km) and high geothermal gradient. The main body of the lithospheric mantle in this area is relatively young, mainly composed of lherzolite with full major elements and depleted Sr–Nd isotope composition, which has characteristics similar to the "oceanic" lithospheric mantle (Fig. 3.11). These conclusions are further supported by the research results of geophysics, palaeo-temperature and Re-Os isotope data. On the one hand, this evidence proves that the lithosphere of the eastern NCC has undergone large-scale thinning since the Phanerozoic; on the other hand, it fully shows that the material composition and properties of the lithospheric mantle in the eastern NCC have undergone significant changes.

The transition process of lithospheric composition under the NCC results from the interaction between peridotite/granulite and melt from different sources, which is the essence of the destruction of the North China Craton. There is obvious Sr–Nd and Li-Fe-Mg isotope disequilibrium among minerals of mantle peridotite, which is caused by the interaction between peridotite and melt. This further proves that there is indeed the interaction between melt/fluid and peridotite in the mantle. The relatively low Os content and high Os isotope ratio of mantle peridotites in the NCC

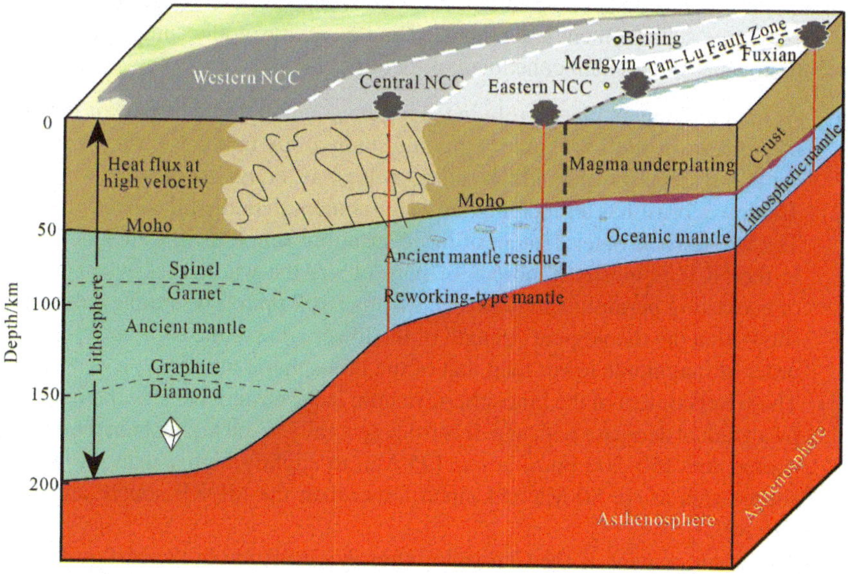

Fig. 3.11 Cenozoic lithospheric section of the North China Craton. The Cenozoic lithosphere of the North China Craton is thin as a whole. The thickness of the lithosphere in the eastern part of the NCC is only 60–80 km, in the central part (Taihang Mountain area) approximately 100 km, and in the western part (Ordos block) 150–200 km

are also the result of peridotite–melt interaction, that is, mantle metasomatism can lead to changes in Os content and isotopic composition of the lithospheric mantle.

Therefore, the lithospheric mantle of the NCC experienced multiphase and multi-stage melt reworking processes, resulting in great changes in its composition and properties. The systematic experimental study of the remaining lithospheric mantle samples further proves that the peridotite–melt interaction is the direct cause of the changes in the composition and properties of the NCC lithospheric mantle, which is also the general feature of the evolution of global cratons.

3.3 *Influence of Fluids on the Stability of the Continental Lithosphere*

The stability of the continental lithosphere depends to a large extent on its rheological strength. Since fluids have a significant influence on the rheological strength of minerals/rocks, understanding the occurrence, content and distribution characteristics of fluids in the continental lithosphere is necessary for studying their stability. Lithospheric fluids are mainly composed of C, H, O and S, and are characterized by high H contents.

Taking the Meso-Cenozoic basalts and the peridotite and granulite xenoliths of the NCC as the research objects, we obtained the following key observations by analyzing the occurrence state, content and hydrogen and oxygen isotopic composition of the structural water in the main constituent minerals.

(1) Although the main phases (olivine, pyroxene, garnet and feldspar) of the deep lithosphere in the NCC are "nominally anhydrous minerals", they generally contain structural water, which can be as high as thousands of ppm.
(2) The lower crust has a much higher water content than the lithospheric mantle. The whole-rock water content of basic granulite xenoliths in Cenozoic basalts in the NCC is mostly hundreds of ppm, while the whole-rock water content of coexisting peridotite xenoliths is less than 100 ppm. The simple calculation shows that the rheological strength of the lower crust of the NCC is at least one order of magnitude lower than that of the lithospheric mantle.
(3) The water content of the lithospheric mantle in the eastern NCC is characterized by a gradual decrease with age. It is from generally much higher than the MORB source area (50–200 ppm) at *ca.* 125 Ma, to slightly higher than the MORB source area at *ca.* 80 Ma and much lower than the MORB source area after <40 Ma.
(4) The water content of the Cenozoic lithospheric mantle in the eastern NCC is generally much lower than that in the MORB source area. Only in weak zones (such as the Tan–Lu Fault Zone) can materials with similar water content in the MORB source area appear.

subducted oceanic crust components have been found in the e areas in North and Northeast China since 90 Ma, which altered basalt and a small number of sediments in the upper crust, as well as the accumulated gabbro and ultramafic rocks f the subducted oceanic crust, indicating that the completely crust was involved in magmatism in this period.

of the lithospheric mantle in the eastern NCC decreases , highlighting the role of water in the process of lithospheric existence of oceanic plate subduction. During the subduction cific Plate, a large amount of water was released into the e of the NCC, resulting in strong hydration, which destabilized tion system in the eastern NCC. The overlying lithosphere was nelted and weakened, which led to lithospheric thinning and .

the Paleo-Pacific Plate is the first-order external driving force tion of the NCC. The dynamic mechanism of NCC destruc- ed as follows: during the late Mesozoic, the Paleo-Pacific Plate ted during continuous subduction underneath the East Asian nained dehydrated in the mantle transition zone. Therefore, the c mantle (large mantle wedge) underwent dynamic processes g and unsteady flow, resulting in the destruction of the eastern ne, the periods of large-scale magmatism, regional extension ralization reflecting craton destruction correspond to the time agnation of the Paleo-Pacific Plate underneath the East Asian Mesozoic. From the perspective of space, the north–south gravity an–Lu Fault Zone and a series of extensional structures in the allel to the Paleo-Pacific subduction zone of the late Mesozoic phenomena. The subduction process of the Paleo-Pacific Plate c had a strong impact on the structure and physical properties (from the surface to the upper mantle and the mantle transi- in the highly uneven shape of the discontinuity surface of the is leads to an unstable mantle flow system under the NCC and ion and melting of the upper mantle or upward flow of mantle her hand, the unstable mantle flow system and the subduction Plate also caused back-arc extension. The interaction of such a d and an unsteady mantle flow system led to the destruction of the numerical simulation results in Fig. 3.13 shows the characteristics dehydration.

ion shows that the NCC destruction in the late Mesozoic was e active global tectonic background at that time. In the background arming, the rapid subduction of the Pacific Plate during this period mergence of unsteady mantle flow in the subduction zone and The addition of a large amount of melt/fluid greatly enhanced of the overlying lithospheric mantle, and the lithospheric mantle

(5) The hydrogen and oxygen isotope analysis of minerals in peridotite xenoliths from the Cenozoic basalts in the eastern NCC shows δD, $\delta^{18}O$ and nonequilibrium fractionation of coexisting minerals are different from the normal mantle, which indicates that the lithospheric mantle was affected by fluids from the subducted oceanic crust.

These observations, combined with the research results of petrology and geochemistry, indicate that fluid has an important impact on the stability of the lithosphere in the NCC. The *ca.* 125 Ma is the peak of the NCC destruction. At this time, the water content of the lithospheric mantle is much higher than that of the MORB source area. This confirms a conjecture of the scientific community over the years that the destruction of stable cratons is closely related to the significant reduction in strength caused by significant hydration. The hydration of the Mesozoic lithospheric mantle in the NCC should be caused by multiple plate subduction around the craton during the late Paleozoic to Mesozoic, which is consistent with the information of hydrogen and oxygen isotopes. The lithospheric mantle of the NCC at <40 Ma has a water content much lower than that of the MORB source area, which is consistent with its reentry from an active state to a stable state, indicating that the stability of the continental lithosphere is related to the high strength caused by strong dehydration. The lithospheric mantle of the NCC in this period may be the residual after thinning. The samples with water contents similar to that of the MORB source area in the weak zone may be newly accreted materials.

The water content in the lithospheric mantle of the NCC decreases gradually from *ca.* 125 Ma to *ca.* 80 Ma and then to <40 Ma, which shows that the change in water content is caused by thermal disturbance in lithospheric thinning. Therefore, the process of lithospheric thinning should conform to the long-term mechanism rather than the short-term mechanism corresponding to the overall delamination. This is consistent with the significant difference in water content between the lower crust and the lithospheric mantle. The difference in rheological strength indicates that the lower crust and the lithospheric mantle are mechanically decoupled and cannot be delaminated together.

3.4 Structure and State of the Crust–Upper Mantle

The structure and state of the crust–upper mantle record the textural response of craton evolution, which is an indispensable basis for understanding the destruction mode and dynamic action of cratons. The seismic exploration of the earth's internal structure in the NCC and its adjacent areas has measured the velocity anomaly and anisotropy of the upper mantle and the shape of the mantle transition zone, which provides textural evidence of local unstable mantle flow in the upper mantle of the NCC.

The new data from intensive sampling and the high-resolution tomography obtained by the finite-frequency seismic tomography have proven that the seismic waves in the shallow upper mantle under the eastern NCC are dominated by low-velocity anomalies, reflecting the characteristics of the hot and thin lithospheric mantle. The seismic wave velocity in the western NCC is obviously high, and the high-velocity block extends below 300 km in depth, corresponding to the thick lithospheric mantle root. The depth from 80 km to the mantle transition zone in the central part of the NCC was detected. The seismic wave at this location shows an obvious low-velocity anomaly with a north–south difference, and the amplitude of the anomaly is stronger than that in the eastern NCC. The low-velocity zone represents strong mantle movement since the Cenozoic, which may originate from the ascending thermal material of the mantle transition zone and even the lower mantle.

Using the SKS wave splitting measurements, more than 2000 effective shear wave splitting observation data were obtained, indicating the upper mantle anisotropy and its reflected upper mantle deformation characteristics in the NCC and its adjacent areas. In the high- to low-velocity anomaly junction area in the central part of the NCC, rapid spatial variation in shear wave splitting parameters was observed. It reflects the effect of the interaction between the asthenosphere and the lithosphere. The combined results of tomography and shear wave splitting indicate that after the upwelling asthenosphere reaches the bottom of the lithosphere, the main body of asthenosphere turns to horizontal flow due to interdiction.

Based on the common conversion point stacking of the receiver function, images of the discontinuity depth and thickness distribution of the mantle transition zone at 410 km and 660 km under the NCC were obtained. The state of the mantle transition zone in the NCC is not uniform in the transverse direction but also changes in the vertical direction. The undulating characteristics of the 410 km discontinuity and its reflected thermal state are consistent with the east–west differences in Meso-Cenozoic magmatic tectonics and craton destruction. The undulating shape of the 660 km discontinuity, the thickness change in the mantle transition zone and its corresponding low-temperature environment represent the retention characteristics of the subducted Pacific Plate at the bottom of the upper mantle in this region. Near western Bohai Bay, the mantle transition zone is relatively thin, and the seismic wave velocity is relatively low, which is obviously different from the regional characteristics. This indicates that the shape of the retained plate is complex and partially broken, and there may be plate fragments passing through the mantle transition zone and entering the lower mantle.

The spatial relationship between the lithosphere and the asthenosphere, the heterogeneity of the upper mantle velocity and the mantle transition zone, and the complex mantle deformation mode provide structural evidence of local unstable mantle flow in the upper mantle of the NCC.

3.5 Dynamic Mec Destruction

Understanding the dynai
understanding its causes
continental evolution. S
have carried out in-deptl
information and made ou

A large number of stud
were affected by subducti
the Paleo-Asian Ocean in
active continental margin.
northward under the NCC
these two subductions did

The impact of Paleo-Pac
It is the key to the overall lc
evidence of the Paleo-Pacifi
lithosphere includes the foll

(1) The study of shallow s
of the eastern NCC cha
change rule is consistei
oceanic plate in the Paci
eastern NCC occurs uni
sion, and the movement
extensional direction of tl
(2) Deep geophysical explora
Plate is inserted under th
Mountains, making the lat
oceanic plate.

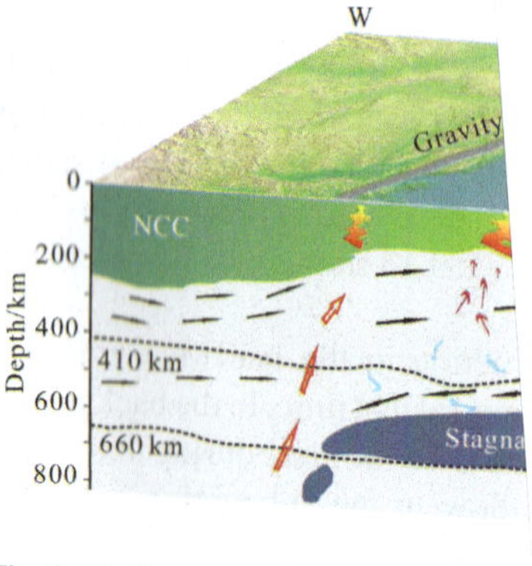

Fig. 3.12 Deep process and dynamic me

(3) A large number of
basic magma sour
may come from th
part of the oceanic
in the lower part o
subducted oceanic
(4) The water conten
gradually with ag
destruction and th
of the western P
lithospheric mant
the mantle convec
metasomatized, r
craton destructio

The subduction of
leading to the destruc
tion can be summariz
(Izanagi Plate) retrea
continent and then re
overlying lithospheri
such as partial meltin
NCC. In terms of ti
and large-scale min
of subduction and s
continent in the late
gradient zone, the T
eastern NCC are pa
and have associated
in the late Mesozoi
of the eastern NCC
tion zone), resultin
transition zone. Th
causes decompress
material. On the o
of the Paleo-Pacifi
dynamic backgrou
eastern NCC. The
of plate subductio

Global correla
closely related to t
of global mantle w
exacerbated the
its adjacent areas
the metasomatism

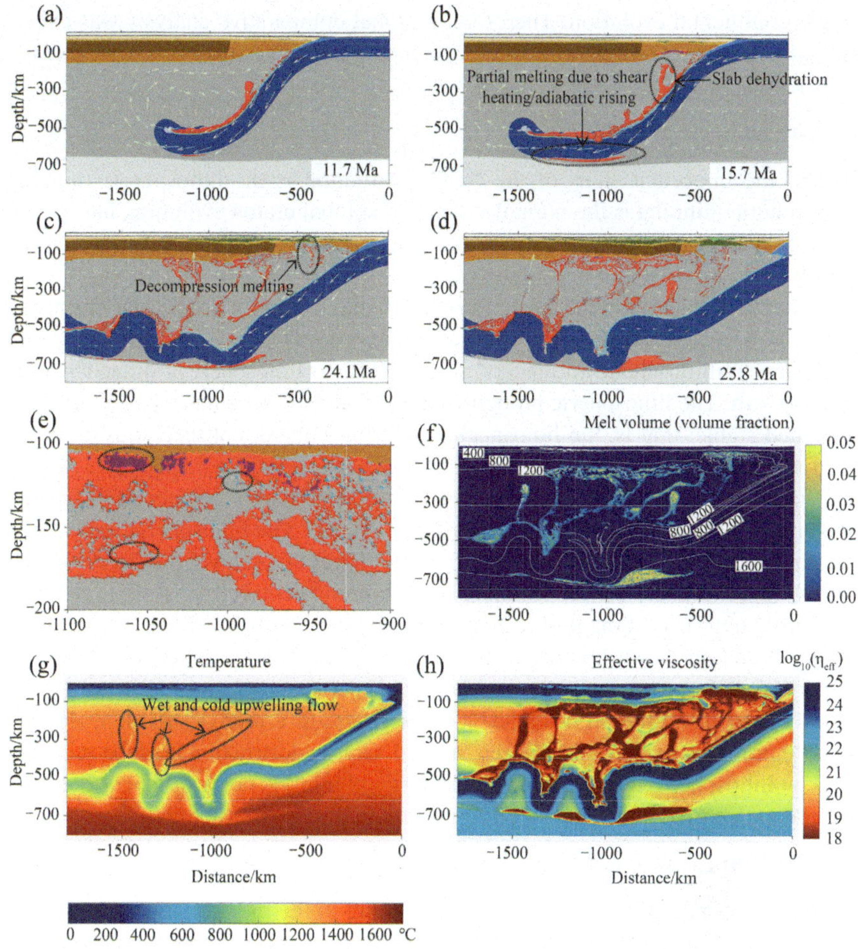

Fig. 3.13 Numerical simulation of ocean subduction and craton destruction process

was significantly weakened. At the same time, the unsteady mantle flow further promoted the tectonic activation of the mechanical weak zone within the craton, thus accelerating the destruction of the cratonic lithosphere.

3.6 *Factors of Global Craton Destruction*

One of the key achievements of the Plan is the discovery of the destruction of the eastern NCC caused by the subduction of the western Pacific Plate. Is the plate subduction also the triggering factor for other craton destruction in the world? Is there a universal law? This discovery is of great significance for understanding the global continental evolution. Therefore, a global comparative analysis was carried out in the implementation of the Plan, which defined the key factors of global craton destruction.

1. **Lithospheric Thinning**

One of the important signs of the NCC destruction is the lithospheric thinning. Lithospheric thinning is the normal state of the global craton evolution, but it is not equivalent to craton destruction. Cratonic lithospheric thinning is common in the world, but many have not been destroyed.

In the stable stage after the formation of the Indian Craton, a variety of nonorogenic magmatic rocks developed. The Deccan basalt erupted over a large area (approximately one million square kilometers) at *ca.* 65 Ma. Its lithosphere is thinned by more than 100 km. The lithospheric properties of the Indian Craton have not changed and remained stable; that is, the lithosphere has been thinned, but there was no craton destruction. The Siberian Craton is considered to be a typical representative of globally stable cratons. Approximately 50 km of lithospheric thinning occurred between 360 and 220 Ma, but craton destruction did not occur. The lithospheric thickness of the South American Craton decreased from ~200 km to ~125 km during 90–70 Ma; that is, approximately 75 km of lithospheric thinning occurred.

Based on the evolution of the Indian, Siberian and South American cratons, lithospheric thinning is not unique to the NCC. It is a geological phenomenon often developed in the process of craton evolution and belongs to the normal state of continental evolution. Different from the NCC, there was no craton destruction, although these cratonic lithospheres have been thinned. Craton destruction must be accompanied by lithospheric thinning, but lithospheric thinning is not a sufficient condition for craton destruction, nor can it be equivalent to craton destruction.

2. **Thermal Erosion of Mantle Plumes**

Although the mechanism of lithospheric thinning in the Indian, Siberian and South American cratons has not been determined, it is more likely to be the result of thermal erosion from mantle plumes. Deccan mantle plume magmatism may be the main mechanism for lithospheric thinning in the Indian Craton. During the lithospheric thinning of the Siberian Craton, the Yakutsk mantle plume activity at 360 Ma and the Siberian mantle plume activity at *ca.* 250 Ma should be the main reasons for the lithospheric thinning of the Siberian Craton. The period of lithospheric thinning in the South American Craton is also the period of Trindade mantle plume activity. Previous studies have shown that large-scale mantle plume activity can lead to lithospheric thinning through thermal erosion at the bottom of the lithosphere, while small-scale mantle plume activity may not cause effective thinning of the cratonic lithosphere.

The South African Craton is a typical example of mantle plume activity without lithospheric thinning. The Yangtze and Tarim cratons in China were affected by mantle plumes in the Permian, but the two cratons were not destroyed at that time. The formation of early Paleozoic kimberlites in the NCC may be closely related to the mantle plume, but there was no obvious lithospheric thinning or destruction of the NCC at that time.

The above examples show that lithospheric thinning may (but not necessarily) occur when cratons are affected by mantle plume activity or mantle upwelling. However, mantle plume activity is not a sufficient condition for lithospheric thinning. Some cratons in the world have experienced lithospheric thinning under the influence of mantle plumes, but none of them have been destroyed. At present, there is no geological fact of craton destruction caused by mantle plumes in the world.

3. **Oceanic Plate Subduction**

In addition to the NCC, the North American Craton, among the 35 major cratons in the world, is another typical craton suffering destruction. During the Mesozoic, the subduction of the Farallon Plate in the eastern Pacific formed the Cordillera orogenic belt of western North American, which destroyed the western edge of the North American Craton and caused a loss of its stability. The western Brazilian Craton in South America has also become an active continental margin due to the continuous subduction of the oceanic plate. It has undergone significant thinning and destruction, resulting in the famous Andean Plateau and related Meso-Cenozoic magmatism and mineralization. Global comparison also shows that oceanic plate subduction is the key factor leading to craton destruction. Lithospheric thinning often occurs in other cratons around the world, but most of them are not accompanied by cratonic destruction. Only when it is affected by the subduction of the oceanic plate can craton destruction occur.

3.7 *Craton Destruction Theory*

A large number of observation experiments and multidisciplinary comprehensive studies have been carried out in the Plan, which shows that the NCC destruction is not only manifested in large-area magmatic activity and tectonic deformation within the crust. The essence of the destruction of the NCC is the loss of craton stability, and the reason is the fundamental change in the composition and properties of its deep lithosphere. This understanding reveals the scientific connotation of the NCC evolution and highlights the essence of the evolution of the craton. Through global comparison and research, it is found that lithospheric thinning is a common phenomenon in global craton evolution, but craton destruction mainly occurs in the continental area adjacent to the plate subduction boundary. The interaction of the subduction plate, the lithospheric mantle and the asthenospheric mantle is the key to the destruction of cratons. Based on the Plan, the craton destruction theory is established and has global applicability. It realized a major breakthrough in theory, and developed the theory of plate tectonics.

As an important geodynamic process, craton destruction has important guiding significance for the formation and evolution of continents. In traditional continental geological research, geologists are extremely concerned about why the continent can be preserved for a long time. The discovery of craton destruction push people into thinking about why the continent was destroyed. The eastern part of the NCC is a typical region suffering craton destruction, but craton destruction is not unique to the NCC. The geodynamic process of craton destruction has also occurred on other continents in the world. The destruction of the western North American Craton is a typical example. Continents amalgamate through collision and tend to be stable through cratonization, but this is not the end of continental evolution. When subjected to the subduction of the surrounding oceanic plate, the craton could be destroyed. When the deep mantle returns to its normal state, the upper continent tends to stabilize and completes a new cratonization process. Therefore, the long-term stable continental craton will lose its stability due to destruction, and craton destruction is also an important link to continental evolution.

4 Shallow Effects of Craton Destruction

The shallow effects of the NCC destruction, especially the resource and biological effects, are related not only to the in-depth understanding of earth system sciences but also to Chinese sustainable development and national strategic needs. The Plan has also made some important achievements in the following aspects.

4.1 Large-Scale Gold Mineralization

Gold is one of the national strategic resources and scarce minerals. North China is an important gold production base in China, and the proven gold reserves account for approximately 50% of the country. The large and superlarge gold deposits in the NCC are mainly distributed in the eastern and central blocks (craton destruction area) and are composed of two eastern and western gold metallogenic belts extending near NNE–SSW (Fig. 3.14). The eastern gold metallogenic belt is represented by the Jiaodong, Liaodong and Jinan gold ore-concentrating areas; the western gold metallogenic belt mainly includes gold ore-concentrating areas such as the Xiaoqinling–Xiong'ershan, the middle part of Taihang Mountain and northern Hebei–eastern Hebei areas.

The large-scale gold mineralization in the NCC occurred in the Early Cretaceous, which is consistent with the peak period of craton destruction. The mineralization has the following main characteristics: (i) Gold and other metallogenic materials are multiple sources, directly from magmatic hydrothermal solution or extracted from Precambrian metamorphic rocks by crustal fluid; (ii) the migration and enrichment of gold are controlled by a unified fluid system; (iii) the mineralization is explosive, and the metallogenic age is concentrated in 135–115 Ma; (iv) the occurrence and distribution of ore bodies are obviously controlled by secondary faults, and most of them have a close spatial relationship with late Mesozoic rock bodies; and (v) under a strong extensional tectonic background, normal faults and metamorphic core complexes are the main ore-controlling structures. These geological and

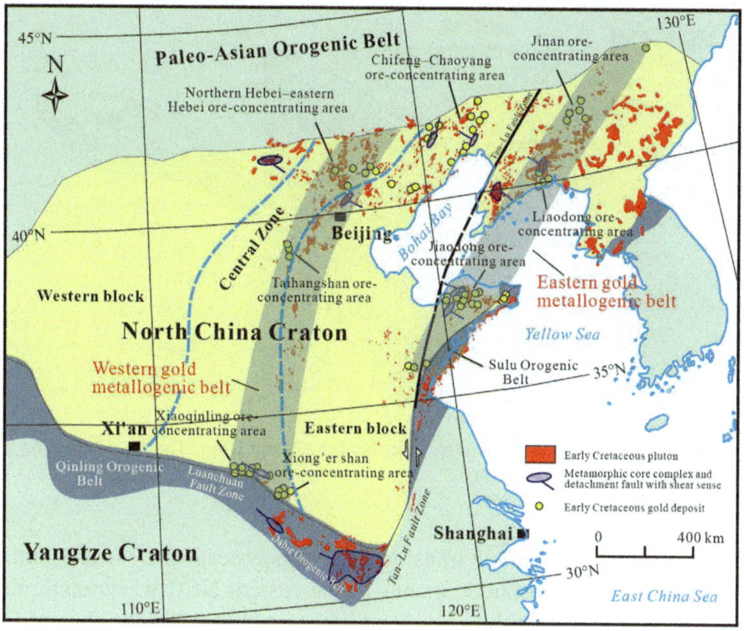

Fig. 3.14 Distribution of early cretaceous gold metallogenic belt in the North China Craton

mineralization characteristics of the Early Cretaceous gold deposits are obviously different from those of vein-type gold deposits in other cratons in the world, and it is difficult to interpret them with the internationally popular metallogenic theory of orogenic gold deposits. Therefore, the Early Cretaceous gold deposits in the NCC do not belong to the "orogen type" but the "craton destruction type". The essential difference between the "craton destruction type" and "orogen type" (compressive background) gold deposits lies in the different metallogenic tectonic background (extensional background) and the source of metallogenic fluid, which are mainly associated with magmatic activities related to craton destruction.

The metallogenic ages of the eastern and western gold metallogenic belts in the NCC are concentrated at 120 and 130 Ma, respectively. The difference in gold mineralization age is caused by the retreat of the western Pacific Plate subduction and the migration of the subduction zone in the Early Cretaceous. The leading edge of the subducted Pacific Plate reached the western margin of the eastern NCC at approximately 130 Ma and stagnated in the mantle transition zone due to the retreat of the plate. The stagnation of subducted plate in the mantle transition zone leads to unstable flow in the upper mantle. This unstable flow causes a sharp increase in the contents of melt and fluid in the lithospheric mantle and gradually transforms it into a strongly metasomatic enriched mantle. The enriched lithospheric mantle interacted with the lower crust to form a fluid rich in gold. These ore-forming fluids rose along the structural weak zone of the crust to form gold deposits in the western gold metallogenic belt. With the eastward retreat of the oceanic plate, the subducted plate

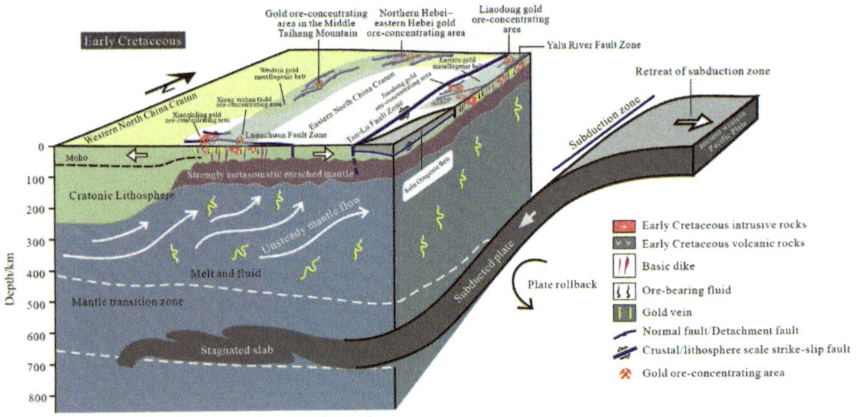

Fig. 3.15 Relationship between the deep process of the North China Craton destruction and gold mineralization in the early Cretaceous

began to stagnate in the mantle transition zone underneath the eastern gold metallogenic belt, and the lithospheric mantle in the eastern NCC was transformed into a strongly metasomatic enriched mantle at approximately 120 Ma. The ore-forming fluid formed by the interaction between melt/fluid (formed by mantle melting) and lower crust rose along major fault zones and their secondary faults, such as the Tan–Lu and Yalu River faults, forming the eastern gold metallogenic belt. This shows that the NCC destruction controlled the gold mineralization (Fig. 3.15).

4.2 Large-Scale Molybdenum Mineralization

During the destruction of the NCC, large-scale molybdenum mineralization also occurred in the southern and northern margins of the NCC. The Qinling molybdenum ore belt on the southern margin of the NCC trends NW–SE, which is the region with the most abundant molybdenum ore resources in China, and the proven molybdenum metal reserves are approximately 8.9 million tons. These results show that the Qinling molybdenum ore belt is the largest in the world. The molybdenum mineralization in this belt is closely related to Yanshanian magmatism, and the mineralization styles are mainly porphyry, porphyry-skarn composite, quartz vein and carbonate vein types. The metallogenic time of the molybdenum mineralization is 210–110 Ma, and the mineralization in the Early Cretaceous is the strongest.

The Yanliao molybdenum ore belt in the northern margin of the NCC trends E–W and extends for more than 1000 km. These molybdenum deposits are also closely related to Yanshanian magmatism. The mineralization styles are mainly porphyry, skarn and quartz vein types. There are three main mineralization ages in this belt, namely, 248–225, 190–145 and 145–134 Ma, and the latter (Early Cretaceous) has the largest mineralization scale.

The formation age of molybdenum deposits in the NCC includes three periods. Molybdenum mineralization began in the Triassic and peaked in the Early Cretaceous. Large-scale molybdenum mineralization occurred during craton destruction. In the Triassic, the molybdenum mineralization occurred outside the southern and northern margins of the NCC, which was related to the evolution of the Paleo-Asian Ocean and the Paleo-Tethys Ocean. In the Jurassic, the NCC turned the circum-Pacific tectonic domain, and some large molybdenum deposits appeared. In the Early Cretaceous, the crust–mantle interaction was significantly enhanced due to the retreat of the Paleo-Pacific Plate, resulting in large-scale molybdenum mineralization, including the formation of superlarge molybdenum deposits.

4.3 *Impact of Craton Destruction on Continental Oil and Gas Distribution*

The NCC destruction also affects the distribution of continental oil and gas. The Ordos Basin in the western NCC is rich in oil and gas of the early Mesozoic age, which developed under a stable and undamaged craton setting. In the eastern part of the NCC, there is no good oil and gas prospect in the Hefei, Zhoukou and Jiaolai basins developed in the Early Cretaceous (peak destruction period), indicating an unfavorable impact of craton destruction on oil and gas formation and preservation. No early Mesozoic oil or gas has been found in the Hefei Basin. During the Paleogene, strong extensional activity occurred again in the eastern NCC, forming the large-scale petroliferous Bohai Bay Basin, which is an important oil and gas accumulation area in China. This large basin appeared in the strongest part of the lithosphere thinning, and lithospheric thermal subsidence occurred after basin formation. Compared with the large terrestrial basin on the stable craton, the Bohai Bay Basin has obvious heterogeneity and complexity of internal evolution and oil and gas accumulation, which also reflects the significant influence of the strong deformation of the craton on reservoir formation.

5 Development of New Observation Technology

Guided and driven by scientific issues, a first-class experimental and observation technology platform has been built. In the field of geochemical experimental research, a series of in-situ microdomain chronology and isotope analysis methods have been independently developed, and these pioneering methods and technologies have been widely applied to the study of the NCC destruction and global continental tectonics. The analytical methods and technical processes established in the project have been adopted by many international laboratories. According to the statistics of ISI Geosciences, at present, the laser in-situ Sr–Nd-Hf isotope data published by the

Multicollector-Inductively Coupled Plasma Mass Spectrum laboratory of the Institute of Geology and Geophysics, Chinese Academy of Sciences account for approximately half of the whole international academic community, providing a high-quality technical platform for geoscientists in China and abroad. The new methods, such as nontraditional stable isotopes developed by this project, have opened a new era for Western scientists to learn and apply the "Chinese methods".

A series of in-situ microdomain dating methods for accessory minerals has been developed, and the types of accessory minerals that can be dated are the most in similar laboratories in the world. It is an important tool for determining the age of geological bodies lacking zircon and baddeleyite. At present, the minerals that can be measured include plagioclase, calcite, sphene, apatite, perovskite, monazite, bastnaesite and pyrochlore.

Kimberlite is the parent magma of diamond, which has a deep origin and contains much information about the composition and evolution of the lithosphere. Kimberlite contains a large number of inclusions and has strong weathering and alteration, which lead to great difficulties in the determination of kimberlite age and initial magma composition. To solve this problem, the in-situ isotopic analysis technology of perovskite in kimberlites has been developed. The isotopic composition of the Paleozoic lithospheric mantle of the NCC was accurately determined by this technique. It solved the problems of kimberlite age and genesis.

To accurately determine the time of NCC destruction and trace the process of the property transformation of the lithospheric mantle, the Cameca 1280 Ion Probe and Nano Ion Probe have been introduced in this project. Through the development of a series of element and isotope analysis methods and standard samples, high precision, high sensitivity, high spatial resolution, efficient isotope dating and tracer analysis have been achieved (Fig. 3.16). The Cameca 1280 Ion Probe is used for high-precision microdomain analysis with a spatial resolution of 5–20 mm. It can accurately determine the content and isotopic composition of various elements in geological samples. Therefore, it is determined that the peak time of the destruction of the North China Craton was the Early Cretaceous (*ca.* 125 Ma). Nano ion probes are mainly used for high spatial resolution microdomain analysis with a spatial resolution of <5 μm–nm, which solves the problem of the in-situ determination of microdomain samples (Fig. 3.17). Samples with fine particles cannot be accurately determined by the laser technology or the Cameca 1280 Ion Probe.

It is the first in the world to carry out online determination, connecting the laser system with two plasma mass spectrometers to realize the simultaneous determination of age, trace elements and isotopes, which greatly improves the analysis efficiency and can obtain the matching chronological and geochemical information of the same part of the sample. For example, a new online simultaneous in-situ analysis method of U–Pb age, Hf isotope and trace elements of zircon and baddeleyite has been established, which realizes obtaining a variety of isotope and element composition information when sampling at the same part. This method has been widely used in many research fields, such as petrological genesis, continental crust formation and orogenic belt evolution. This technology realizes the efficient, rapid and accurate

Fig. 3.16 Cameca 1280 ion probe

Fig. 3.17 Researchers introduce the characteristics and applications of Nano ion probes to visiting scholars

Fig. 3.18 Chinese and foreign scholars conducted a joint field scientific investigation on the NCC (D. McKenzie, Crawford prize winner and academician of the Royal Society, stands in the middle of the picture; Photo location: Hannuoba, Hebei)

determination of a large number of samples and attracts a large number of foreign first-class scholars to cooperate in research (Fig. 3.18).

The establishment of the first-class experimental research platform not only provides high-level in-situ element and isotope microdomain analysis technical support for geoscientists but also improves the international academic competitiveness of China's solid geoscience research.

Outlook

Craton destruction belongs to the continental evolution. Continental evolution is an important frontier field of geoscience that cannot be explained by classical plate tectonics theory in many aspects. In recent years, a series of major research plans have been set up to explore the formation and development of continents and try to establish a new theoretical system of continental evolution. The Chinese mainland's geological conditions richly endowed by nature provide a rare natural laboratory for China's geoscience community to achieve breakthroughs in geoscience theory. The long-term accumulation of research in the Chinese mainland has enabled our scientists to have a solid foundation for the creation of new theories of continental evolution. China's growing economic strength and increasing scientific research investment have also created good opportunities for Chinese scientists to carry out collaborative and innovative research, develop plate tectonics theory and achieve a breakthrough in continental evolution. Seizing this rare opportunity and realizing theoretical breakthroughs, China's geosciences will lead the international geoscience development in the field of continental evolution.

The implementation of the Plan has gathered outstanding scientists in the field of geoscience, coordinated research around clear scientific objectives, greatly improved China's scientific research ability in the field of geoscience, and made outstanding contributions to China's becoming an international geoscience power. The research results of the Plan not only have an important impact on the current study of continental evolution but are also expected to make breakthroughs in the following geoscience research fields when we look forward to the future.

© Zhejiang University Press 2023
R. Zhu (ed.), *Research on Destruction of the North China Craton*, Reports of China's Basic Research, https://doi.org/10.1007/978-981-99-6046-0_4

1 Tethys Geodynamic System

The theory of plate tectonics was founded in the late 1960s [15] and then became a unified theoretical framework for all branches of earth sciences. Its importance is comparable to the significance of the relativity theory and quantum mechanics to physics and the DNA to life sciences. It is one of the most important scientific discoveries in the twentieth century. The core of plate tectonics theory is that plates are rigid, and their motion law conforms to Euler's theorem. Plate theory has been born for 50 years, but geoscientists have been exploring how the earth's interior works. The American Association for the Advancement of Sciences also listed it as one of the top ten scientific issues in the twenty-first century [18, 27]. The important enlightenment of the implementation of the Plan is that geoscience research must have a global perspective.

From the perspective of exploring the overall operation of the earth, the collisional orogenic belt is another kind of tectonic unit as important as the craton on the earth. The Tethys orogenic belt is an arc–continent collision orogenic belt gradually formed during the extinction of the Tethys Ocean at different times, which is a natural laboratory for exploring the dynamic process of the earth's interior. The Tethys tectonic domain has completely preserved the geological records of different processes of inter plate movement of continental breakup (Red Sea to East African Rift), ocean–ocean subduction (Tonga Trench), ocean–continent subduction (Sumatra), arc–continent collision (northern Australia), continental collision (Zagros, Iran) and continental deep subduction (ALPS) [23] (Fig. 4.1). The Tethys tectonic domain also contains many important geological phenomena that cannot be explained by classical plate tectonic theory, such as intracontinental deformation and plateau uplift (Tibet Plateau) [21], mineralization of continental collision [5], and ophiolite-type diamond [31]. Among the 9 international comparison plans with the theme of geodynamics established by the United Nations Education Scientific and Cultural Organization (UNESCO) and the International Union of Geological Sciences (IUGS), the themes of 7 plans are closely related to the dynamic evolution of the Tethys.

The primary evolution feature of the Tethys tectonic domain is the continuous northward drift and convergence of continental blocks to the south of the Tethys Ocean [25] (Fig. 4.2). For example, the Gondwana continent in the Southern Hemisphere and the Laurasia continent in the Northern Hemisphere converged at ~300 Ma to form the Pangaea Supercontinent, and the middle part formed the Paleo-Tethys with the eastward opening. Subsequently, the northern margin of the Gondwana continent was broken up to form the Kimeri continent, and a Neo-Tethys Ocean was generated on its southern side. The Kimeri continent was collaged northward to the south of the Laurasia continent at 200 Ma, resulting in the closure of the Paleo-Tethys Ocean. The Pangaea supercontinent began to break up at 150 Ma, forming a series of continental blocks drifting northward, such as the African, Arab, Indian and Australian plates. These separated plates collided with the Eurasian continent at 50 Ma, closing the Neo-Tethys Ocean and forming the Alps-Carpathian orogenic

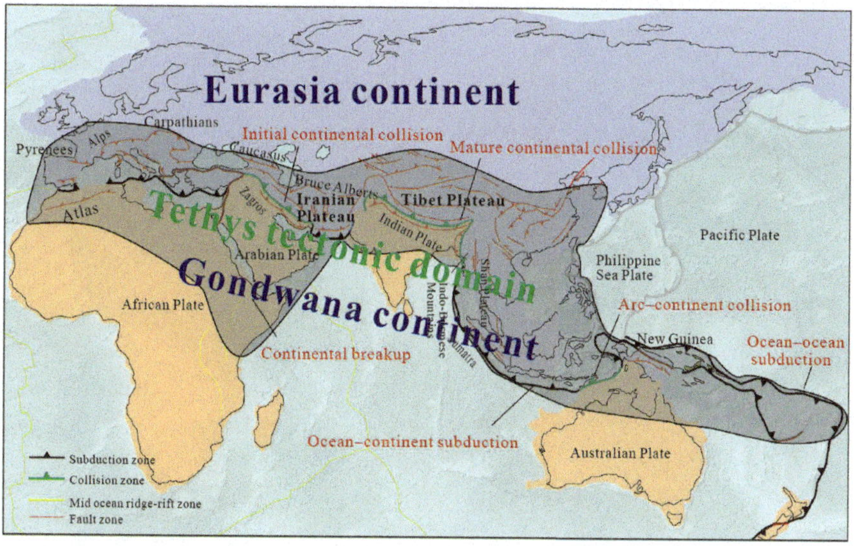

Fig. 4.1 Schematic diagram of the Tethys tectonic domain

belt, Turkey-Iran Plateau and Tibetan Plateau [25]. At present, the puzzling question is why do many land blocks with different properties in the Tethys tectonic domain converge unidirectionally with Eurasia to the north? What is the deep dynamic mechanism that gradually tore apart from the southern hemisphere continent and then drove the broken blocks to collage the southern edge of Eurasia? These problems essentially involve the fundamental scientific problem of the plate tectonic driving force. The classical plate tectonic theory still cannot explain many phenomena in the Tethys tectonic domain. Therefore, it is necessary to develop new geoscience theories to reasonably explain some global tectonic problems.

The Tethys tectonic domain is one of the three major metallogenic domains in the world, namely, the Tethys metallogenic domain [20] (Fig. 4.3). Although the Tethys tectonic domain has experienced long-term oceanic plate subduction and large-scale magmatism, the associated ore deposits are very rare. The most important porphyry deposits in the Tethys tectonic domain were basically formed after continental collision. What kind of geodynamic process causes such a unique metallogenic effect in the Tethys domain? This is a major theoretical issue with great concern to the global ore deposit academic community. The Tethys tectonic domain is also the most important oil and gas resource accumulation area in the world, and the remaining recoverable reserves of oil and gas account for approximately 70.6% of the global total [22]. However, the distribution of oil and gas resources in the Tethys tectonic domain is extremely uneven, and most of them are concentrated in the Persian Gulf. Why is the distribution of oil and gas resources so uneven under similar tectonic backgrounds and similar paleo marine environments? Why can the passive continental margin of the Arabian Plate form a series of fold structures, while the northern margin of the

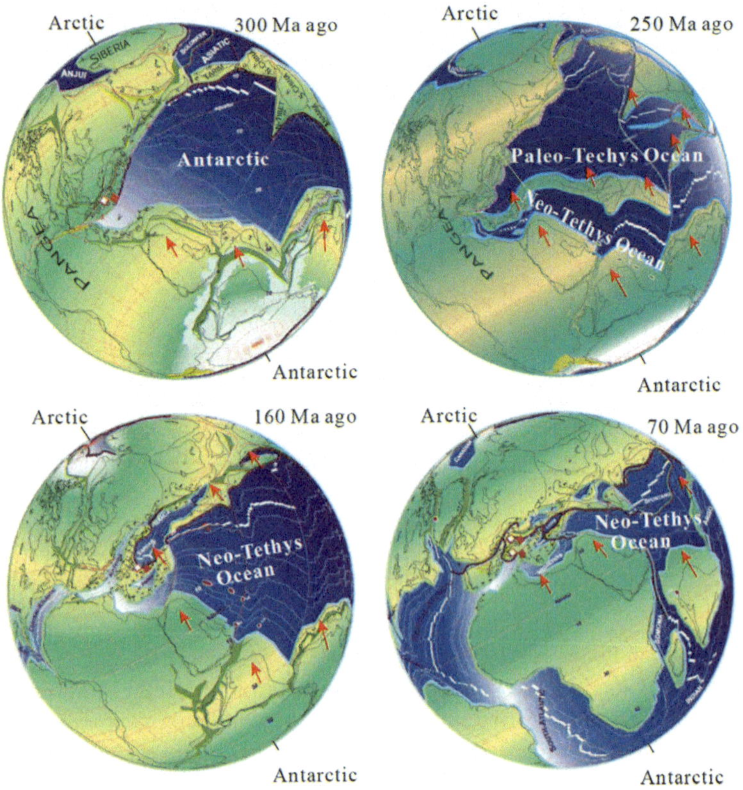

Fig. 4.2 Evolution of the Tethys ocean and unidirectional convergence of continental blocks

Indian Plate, a passive continental margin, caused the towering Himalayas? Detailed anatomy of the dynamic process of many sedimentary basins in the Tethys tectonic domain is an important way to understand the "mystery of the Tethys hydrocarbon".

The evolution of the Tethys tectonic domain is the result of the joint action of shallow and deep lithospheric processes, which deeply affected the earth's surface system and finally established the environmental pattern of the modern earth [9]. The demise of the Tethys Ocean has led to the closure of ribbon oceans in low latitudes. At the same time, the volume of global oceanic basins has increased, and the continental area has expanded with the opening of the ribbon oceans in middle and high latitudes and the expansion of oceanic basins. The above processes led to a gradual change from the salinity-dominated ocean current system to the modern ocean circulation system driven by the earth's temperature difference in the Cretaceous greenhouse earth. The closure of the Neo-Tethys Ocean and resultant continental collision led to the uplift of the Tibetan Plateau, affected the global atmospheric circulation and oceanic circulation, enhanced the efficiency of heat exchange in high and low latitudes, and led to the transformation from "greenhouse earth" to "ice chamber earth". However,

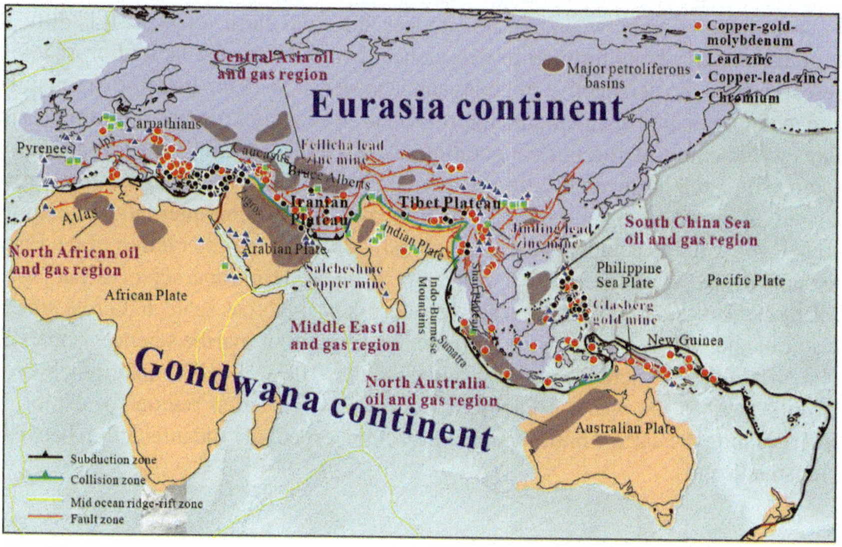

Fig. 4.3 Main metal minerals and petroliferous basins in the Tethys tectonic domain

the specific impact mode and mechanism of the Tethys evolution on major global events are still major unsolved scientific problems.

In short, the Tethys geodynamic system aims to study the opening and closing process of the Tethys Ocean between northern Laurasia and the southern Gondwana continents and the related resource, energy and environmental effects. The in-depth study of the Tethys orogenic belts is a bridge for Chinese geoscientists to go abroad. The Tethys tectonic domain, the most concentrated area of geoscience research and the richest area of energy resources, is highly coincident with the area covered by China's "The Belt and Road Initiative". Through the comprehensive study of the Tethys tectonic domain, it can better serve the strategic needs of national resources and energy and the objectives of economic and social development.

2 Multi-sphere Interaction of the Earth System in the Western Pacific

Earth system sciences take each sphere of the earth as a unified whole and study the laws and control mechanisms of various processes of the earth based on the interaction between the earth's outer atmosphere, hydrosphere, biosphere and the earth's inner crust, mantle, core and other subsystems. However, since the earth system science was put forward more than 30 years ago, there has been no substantive breakthrough in the research of multi-sphere interactions. An excessively detailed division of discipline fields leads to insufficient interdisciplinary exchanges. A typical example is the lack of cross integration between marine sciences and solid earth sciences. It is difficult to make a major breakthrough in geoscience theory due to the lack of cross circle research from the perspective of geodynamic systems. With the increase in scientific and technological investment, the improvement of scientific research conditions and the strengthening of foreign cooperation and exchange, China's marine and earth science research has been in a turning period from quantitative change to qualitative change in recent years, and is struggling to catch up with the international cutting-edge level. The forthcoming research, named "Multi-sphere Interaction of the Earth System in the Western Pacific", will help to integrate marine sciences and solid earth sciences and focus on the research of cross basin, cross scale and cross sphere material and energy exchange to promote the research of marine sciences and earth system sciences to achieve major original results. The well-developed trench–arc–basin system in the western Pacific is an excellent selection region for exploring and revealing the earth's multi-sphere interaction, especially the interaction between fluids and solids.

2.1 Reconstruction of Western Pacific Plate Subduction

The trench–arc–basin system where oceanic and continental plates converge is the entry point for the interdisciplinary intersection of marine sciences and solid earth sciences. The Plan reveals that western Pacific Plate subduction during the late Mesozoic was the driving force of the NCC destruction. However, there is insufficient research on the interaction mechanism between the subduction process of the Paleo-Pacific Plate and the overlying continental lithosphere.

There is an important consensus in geosciences that the destruction of the NCC is directly related to the subduction process of the western Pacific Plate. However, important scientific problems, such as when the western Pacific Plate began to subduct toward the East Asian continent, where the Early Cretaceous western Pacific subduction zone was, when the large mantle wedge in East Asian was formed, and how the late Mesozoic western Pacific subduction plate evolved, have not been well solved. The existing tectonic models suggest that the westward subduction and eastward retreat of the western Pacific Plate led to the strong extension and thinning of the lithosphere of the NCC, while the stagnation and dehydration of the subducted plate in the mantle transition zone led to the partial melting of the upper mantle and large-scale magmatism of the crust. In addition, traces for the subduction of the oceanic plate under the East Asian continent and the stagnation in the mantle transition zone are the result of the subduction of the Cenozoic western Pacific Plate rather than the record of the subduction process of the Mesozoic Paleo-Pacific Plate. Based on geological and paleontological studies, the coastal mountains on the edge of East Asia were 3500–4000 m above sea level in the Late Cretaceous, extending approximately 500 km inland from the coast [3]. Compared with the current situation of the eastern Pacific subduction zone, we believe that the Cretaceous western Pacific subduction zone is located on the continental margin of East Asia, which is closer to the west than the current western Pacific subduction zone. At present, the recovery of the subduction history of the western Pacific Plate is mainly based on the petrology and geochemical analysis of the NCC, the study of crustal deformation and the numerical simulation of global plate motion [17, 24, 32, 36]. Various plate subduction models established before do not truly reveal the detailed subduction history of the western Pacific Plate. We still know little about some basic problems, such as the origin of the ancient western Pacific Plate, the time of initial subduction, the change in subduction direction, the change in subduction angle and subduction rate, and whether there is mid-ocean ridge subduction. Therefore, the reconstruction of the subduction history of the western Pacific Plate will be a very challenging research field for China's geoscience. To make a breakthrough in this research field, it is obvious that we should not only carry out inversion based on the geological, geochemical and geophysical data of the East Asian continent but also carry out various oceanic exploration and geological history analyses of the western Pacific Plate. The direct acquisition of exploration data and the continuous accumulation of key evidence will lay a solid scientific foundation for reconstructing the subduction history of the western Pacific Plate.

It is inferred that the subduction of the western Pacific Plate to the East Asian continental margin began in the Early Jurassic, and the subduction zone is located on the East Asian continental margin. The time when the western Pacific subduction plate stagnated in the mantle transition zone and the history of the large mantle wedge in East Asian are frontiers of geoscience in the future. The large mantle wedge is very different from the small mantle wedge near the trench. Since the distance from the trench is thousands of kilometers, the area involved in the large mantle wedge is generally considered to be an intraplate environment. The large mantle wedge, the stagnated subduction plate in the mantle transition zone and the lower mantle together constitute a unique "sandwich" mantle structure. The time when the subduction plate begins to stagnate in the mantle transition zone is roughly the same as the time when the subduction zone retreated [30]. According to the spatio-temporal migration law of late Mesozoic magmatic rocks on the continental margin of East Asia, it is speculated that the large mantle wedge in East Asia may have been formed in the Early Cretaceous, and the influence of the western Pacific Plate subduction on the NCC may be realized through the migration and exchange of material and energy in the large mantle wedge. The large mantle wedge deep in the continental margin of East Asia not only has an important impact on the NCC destruction but also controls the formation of intraplate basaltic magma in this region. The detailed evolution process of the large mantle wedge in East Asia and its impact on major events, such as the deep carbon cycle, Earth's climate system and terrestrial biological evolution, will also become a frontier issue in earth system sciences.

2.2 A Window for the Study of Marine Systems

The western Pacific Ocean is a unique and complex marine area that has the world's largest ocean warm pool—the western Pacific warm pool, the strong western boundary current—the Kuroshio, active mesoscale vortices and small-scale internal waves and other multiscale motion. It is also the core area of typhoon and the world's main climate mode El Nino-Southern Oscillation (ENSO) and Pacific Decadal Oscillation (PDO). A variety of trench–arc–basin systems and complex ocean–continent patterns and seafloor topography are developed in the western Pacific Ocean (Fig. 4.4), which play an important role in regulating ocean dynamic processes such as ocean–ocean exchange, ocean–continent exchange, sea–air exchange and upper-deep ocean exchange. The westward subduction of the Pacific Plate leads to large-scale magmatic and fluid activities, which can have an important impact on the material circulation and energy exchange between the ocean and deep earth. Through interdisciplinary research, researchers have deeply and systematically studied the impact of the western Pacific Plate movement on the seawater dynamic process and air-sea interaction under the complex ocean–continent pattern, the subduction amalgamation and transformation between the western Pacific Plate and Eurasian, Indo-Australian continents, and the material and energy cycle across spherical layers, which will give birth to a new theory of earth system sciences and drive the development of contemporary earth system sciences. In addition, the western Pacific Ocean is a marine region with the most complex marine dynamic process, frequent marine-meteorological/ geological and seismic disasters, and abundant seabed mineral resources and oil and gas resources. Carrying out multidisciplinary and systematic comprehensive research in the western Pacific Ocean, revealing the multiscale dynamic process of the ocean and the convergence-divergence dynamic process of the continental margin, understanding the multicycle material and energy exchange, transmission process and mechanism, comprehending the mineralization and accumulation mechanism of seabed mineral resources in the western Pacific Ocean, and clarifying the multiscale change law of the marine environment and climate can provide important scientific and technological support for national marine security and disaster prevention and reduction.

2.3 Ocean–Continent Pattern/Seafloor Topography Evolution and Marine Dynamic Process

Plate movement will cause significant changes in seafloor topography and ocean–continent patterns, which plays an important role in regulating marine dynamic processes. For example, the opening and closure of straits and the generation and dissipation of islands may lead to significant adjustment of local and even the whole oceanic basin circulation system [8, 10, 29]. The change in seafloor topography will have a significant impact on the energy cascade and vertical mixing process of

Fig. 4.4 Complex ocean–continent patterns in the western Pacific Ocean and eastern Eurasia

the ocean [13]. Therefore, an accurate understanding of ocean–continent patterns, seafloor topographic characteristics, control factors and regulatory mechanisms of marine dynamic processes is an important prerequisite for comprehending changes in paleoclimate systems and predicting future climate change. This research has just started in the world, and Chinese scientists should pay attention to and actively participate in this frontier research.

Through long-term ocean observation and geological research, European and American scientists have successively found important phenomena, such as mid-ocean ridges, transform faults, subduction zones and submarine magnetic anomaly strips, and then put forward the theory of plate tectonics [11, 15, 16]. However, the classical theory of plate tectonics has not solved two key problems: i. What the driving force of plate movement and the mechanism of plate subduction are. Although it is generally believed that plate motion is controlled by the deep process of the earth, it is still unclear how the earth's interior acts to drive plate motion. Among the 125 scientific questions raised by *Science* on its 125th anniversary, the first question about earth sciences is "how the earth's interior works". ii. How plate subduction starts. The models of plate subduction initiation mechanisms mainly include active and passive subduction models. According to the active subduction model, the density difference between plates causes plates with high density to sink and insert under plates with low density to start subducting [1, 19]. The passive subduction model assumes that the weak zones of the plate bend, break and thrust down under the action of horizontal compressive or transpressive external force to start subduction [4, 28]. Due to the complexity of the initiation of plate subduction itself, there is no unified understanding thus far, which restricts the development of plate tectonic theory.

The key to solving the above two problems is to understand the deep fluid of the earth. Deep fluid is the blood of the earth and the key to understanding the operation law of the earth's interior. Seawater affects the physical properties and chemical composition of oceanic plates through seafloor alteration, sedimentary processes and hydrothermal activity. Plate subduction brings surface fluids to the deep earth, affecting magmatism, plate movement and even the earth evolution. Water entering the earth's interior induces partial melting, leading to the formation of the asthenosphere, which provides the basis for plate movement. Therefore, the mechanism of deep fluid activities, such as the water cycle and carbon cycle, is the core to comprehend the operation law of the earth's interior. The western Pacific subduction zone has a complete trench–arc–basin system and subduction belts of different ages and different types, which is an ideal selection region to study the mechanism of initial plate subduction and improve the theory of plate tectonics. The western Pacific Ocean has always been a focus of attention of geoscientists all over the world, and it is the best natural laboratory to give birth to new theories and ideas in geoscience.

3 Deep Earth Science

The deep process of the earth is the main source power of the operation of the earth system, which not only controls the interaction between the deep spheres but also restricts the evolution of the upper spheres. To trace the internal causes and dynamics of mountains, basins, mineralization, reservoirs and disasters closely related to human survival, we must carefully reveal the deep structure of the earth and study the deep physical–chemical processes and geological processes. The resultant "Deep Earth Science" will show its new vitality. "Deep Earth Science" focuses on the earth's energy conversion, material circulation and dynamics, involving the earth's deep physical structure and state, deep material (chemical) composition and distribution, deep interface behavior and conversion, etc. "Deep Earth Science" mainly involves the following three fields.

3.1 Deep Earth Physical Structure and Main Boundary Layer Properties

The prominent feature of the deep earth is the stratification of structure, which is the most obvious difference between the earth and other planets. From top to bottom, the earth exhibits a Moho at the bottom of the crust, a mid-lithosphere discontinuity (MLD), a lithosphere–asthenosphere boundary (LAB), a mantle transition zone (MTZ) and a core–mantle boundary (CMB). The formation of these interfaces is closely related to mineral phase transformation, chemical composition change and tectonism. The significant changes in mantle structure and material composition at the interface will not only hinder or promote deep processes such as plate subduction and plume upwelling, but also change the mantle convection model and deeply affect the stability and tectonic evolution of the lithosphere. Therefore, it is of great

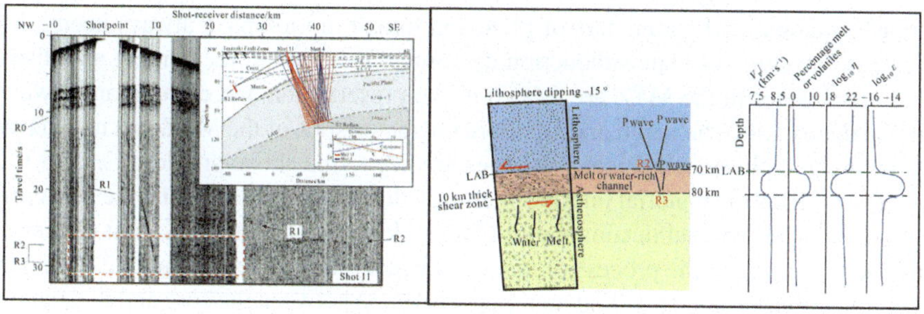

Fig. 4.5 Seismic reflection profile of North Island, New Zealand. **a** There is a low-velocity anomaly zone of approximately 10 km at LAB; **b** Italics decreased by (8 ± 2)%. V_P represents P-wave velocity. ε represents strain rate (s^{-1}) [26]

geodynamic significance to further study the deep structure of the earth, especially the morphological and physicochemical properties of important interfaces.

The mid-lithosphere discontinuity refers to a common interface in the continental lithosphere approximately 100 km deep where the seismic wave velocity decreases significantly with depth. The interface reflects the vertical heterogeneity in the continental lithospheric mantle. It is still unclear how the MLD is formed and whether its generation is related to the craton destruction process.

The lithosphere–asthenosphere boundary is the bottom interface of the rigid lithosphere, and the lower asthenosphere can circulate because it is rich in melt. The seismic reflection profile of North Island of New Zealand reveals that the LAB is a low-velocity anomaly zone approximately 10 km thick, which may be rich in partial melt and fluid. This low viscosity, weak shear region decouples the overlying lithosphere and the underlying mantle flow [26] and promotes plate movement (Fig. 4.5).

The mantle transition zone is an important link connecting the material and energy transmission between the upper mantle and the lower mantle. It is not only the main water storage layer in the inner earth but also an important region for the phase transformation of mantle minerals. The MTZ has a significant effect on the rate, shape and style of plate subduction and mantle plume upwelling. The lower interface of the MTZ is an important factor in determining whether the subduction plate stagnates in the MTZ or directly subducts into the lower mantle, so it is directly related to the scale of mantle convection and the formation of deep geochemical reservoirs. The stagnated slab in the MTZ shows a variety of forms. For example, the scale of the stagnated slab in the MTZ under the continental margin of East Asia can reach thousands of kilometers, forming a unique large mantle wedge, which is obviously different from the small mantle wedge. In the southwestern Pacific and Peru in South America, the subducted plate is directly inserted into the lower mantle and stagnates at a depth of approximately 1000 km. The Central American subducted plate even sank directly to the core–mantle boundary scope. The differences in subducted plate morphology and subduction styles, the reasons for large-scale slab stagnation in the MTZ, and the relationship between mantle rheological properties, mineral phase transformation, and parameters of plate and trench motion have always been a hot issue in the study of plate subduction dynamics. In addition, the content and distribution of water in the MTZ are related to the preservation and circulation of water in the deep earth. The depth and structure characteristics of the MTZ and its internal structure and physicochemical properties are key to the above problems.

The CMB is the material interface between the siliceous mantle and the iron core. It may also be the final destination of plate sinking and the starting point of mantle plume upwelling in the process of mantle convection. At the same time, it is also an important place for energy (and material) exchange between the core and the mantle. The main structures of the CMB include large-scale low-velocity bodies, mantle plumes, ultralow-velocity layers, etc. Due to the complex structure of the mantle bottom, the understanding of the specific geometric shape, spatial distribution and physical parameters of these abnormal bodies is limited, resulting in great controversy in the understanding of the mineral composition and dynamic process of these abnormal bodies. The main scientific problems include the following: Is the physical characteristics and spatial distribution of the ultralow-velocity region at the core–mantle boundary due to different chemical compositions or partial melting? What is the effect of its spatial distribution and structure characteristics on the formation of mantle plumes? What are the causes of the ultralow-velocity region? Is the mantle plume produced by the accumulation of light matter? How do subducted plates reach the lower mantle or core–mantle boundary? The above scientific problems are of great significance for distinguishing and identifying the scale of mantle convection, understanding the chemical heterogeneity in the earth's interior, and clarifying the formation and evolution process of the earth.

3.2 *Composition and Distribution of Deep Earth Materials*

The composition and distribution of deep earth materials involve the study of deep geochemical reservoirs and the composition, structure, phase transformation and physicochemical properties of minerals under high temperature and high pressure. They play an important role on the earth's internal structure and mantle convection, plate subduction, mantle plumes, earthquakes, magmatism and geomagnetic field. The analysis of deep source samples, theoretical calculations and experimental simulations of high temperature and high pressure are three means to explore the operating mechanism of the earth's interior. The mutual verification of these three means is the key to leading this field.

The Mariana Trench is the most typical intra-oceanic island arc belt on the earth, separating two mantle convection systems and deep reservoirs. The eastern side of the Mariana Trench is the Pacific mantle domain, and the western side is the Indian Ocean mantle domain. The Pacific subduction plate lies flat in the mantle transition zone, resulting in stratified convection in the upper and lower mantle. Above the MTZ is the large mantle wedge, which is a large carbon reservoir containing a large amount of recycled materials [12, 30]. The migration of volatiles, such as H, C, S, N and inert gas elements, in the earth's interior is an important link for many deep geological processes, which provides a key clue for an in-depth understanding of plate movement and mantle plume activity. Volatiles can promote mantle melting and convection and control the formation and evolution of the hydrosphere, atmosphere and biosphere. Although the academic community has carried out much research on volatile matter in the earth's interior, there are still many important problems that have not been fundamentally solved, such as the distribution and migration mechanism of C, S and N, the basic law of volatile evolution in the process of crust–mantle interaction, and the effect of volatile matter on the phase transformation and physical properties of existing phases in the earth's interior.

Bridgmanite and ferropericlase are the main minerals in the lower mantle, in which magnesium silicate perovskite accounts for approximately 75% of the total and calcium silicate perovskite accounts for approximately 5% of the total. Therefore, the study of various physical and chemical properties and possible changes in silicate perovskite is the basis and premise for understanding various geological phenomena and processes in the lower mantle. The correlation between structural distortion or phase transformation of silicate perovskite and isomorphic substitute elements can reveal the physicochemical properties and behavior of these substitute elements under high pressure. This is also an effective way to explore important geological phenomena and geological processes, such as the distribution and differentiation of elements in the lower mantle and the chemical heterogeneity of the mantle. The study of impact metamorphic minerals in the crater found that divalent iron in the high-temperature and high-pressure environment of the lower mantle can directly reduce carbonate or CO_2 to diamond in the absence of reducing agents such as iron metal or sulfide [2]. If the above view is correct, it can be inferred that the lower mantle is rich in ferric iron.

Under high-temperature and high-pressure conditions in the lower mantle, highly oxidized FeO_2 can be produced by pressure-induced decomposition and release of free hydrogen from goethite (FeO_2H) [6]. In the presence of iron or iron oxides, the lower mantle water cycle at 1800 km depth turns into a water hydrogenation cycle [7]. If the subducted plate continuously supplies water to the iron reservoir in the core, the core–mantle boundary will release hydrogen, which will become a large hydrogen generator or a source of hydrocarbons and other biochemical components. During the above process, oxygen will be left to form FeO_2 [14], which accumulates in the form of oxygen-rich patches in the CMB (Fig. 4.6).

3.3 *Interaction, Material Circulation and Dynamics of Geospheres*

Plate tectonics and mantle plumes are the main results of deep earth's material movement and geosphere interactions. The former involves the horizontal movement of plates and the subduction of crustal materials into the earth's interior, while the latter

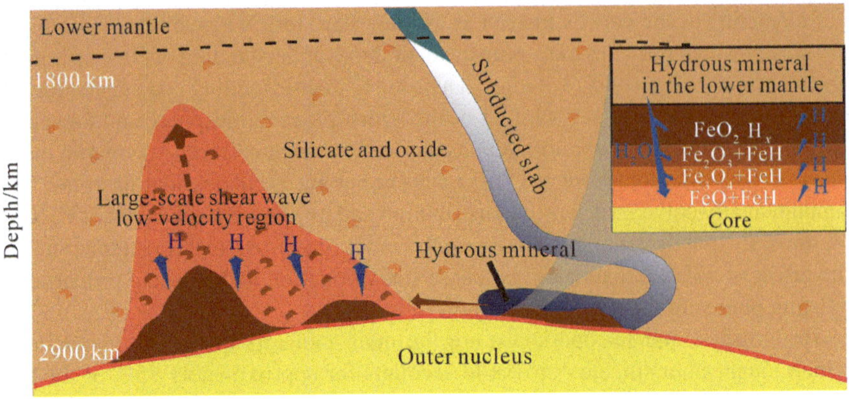

Fig. 4.6 Core–mantle boundary as a huge hydrogen generator and oxygen tank [14]

emphasizes the vertical upward transport of material and energy from the core–mantle boundary. The initiation of global plate tectonics may have been induced by mantle plumes. Mantle plume and plate tectonics constitute an important part of global tectonic theory.

The subduction system at the convergent plate margins is the most active place for material and energy exchange between the shallow and deep parts of the earth. At present, the understanding of how plate subduction began and how the subduction process promoted the formation and differentiation of continental crust is insufficient. Obtaining the mantle melting conditions and source composition characteristics during the initiation of plate subduction will help to understand the mechanism of ocean–ocean subduction. The key to understanding the heterogeneity of island arc magma is to understand the thermal material cycle in the process of oceanic plate subduction, the properties of melt/fluid in the subduction zone, and its magmatic composition and spatio-temporal distribution.

With the passage of geological time, metastable oxygen-rich patches accumulate continuously in the CMB, and finally reach the critical point of mass oxygen explosion. In the long history of the earth, oxygen explosion has directly affected the global environment, biological evolution, earth dynamics and other major processes. The surge in oxygen is a new interpretation for the Great Oxidation Event (GOE), the five mass extinctions and the "Snowball" earth. The explosion of oxygen increased the chemical driving force, which disrupts the steady-state thermally driven mantle convection, leading to the formation and breakup of super mantle plumes and supercontinents. The increase of hydrogen, oxygen and related volatile components can also reduce the melting point of mantle rocks, which leads to large-scale partial melting, forming large igneous provinces and overflow basalts. It may cause catastrophic environmental changes.

4 Craton Destruction and Terrestrial Biological Evolution

Continental tectonics and terrestrial biota are the most influential branches of Chinese solid earth sciences in the world. Chinese scientists have made a series of major research achievements in these two fields [34, 37]. With the support of the Plan, relevant research teams have proposed and gradually improved the new theory of "Craton Destruction", enriched and developed plate tectonics theory [35], and proposed a new ore deposit type of "decratonic gold deposit" [33], which provides a scientific basis for strategic resource replacement. In the field of terrestrial biological evolution, Chinese scientists have improved the "dinosaur origin" hypothesis for birds, and provided key evidence for the arboreal origin hypothesis of bird flight. They put forward a new view that East Asia in the Early Cretaceous was the origin center of several biotic groups rather than a "refuge shelter". However, there is still a lack of a perfect theoretical framework to explain the internal relationship between continental tectonics and the evolution of terrestrial biota. The NCC destruction and the Yanliao and Jehol biotas are two research fields with important scientific significance. Their combination will help to deeply understand the control mechanism of continental tectonics on terrestrial biota and reveal the internal relationship between the surface biosphere and the deep lithosphere in the evolution process.

The earth system includes the earth's interior, the lithosphere, the biosphere, the hydrosphere and the atmosphere, which are related and affect each other. To fully and deeply understand the coupling mechanism of various parts in the earth system, we need to integrate the data of different spheres and disciplines, study the overall behavior of the earth, and establish a complete theory of earth system evolution. However, from a practical point of view, there are many difficulties in the establishment of a complete earth system science, which is mainly reflected in the mutual closure between different branches of disciplines, and it is difficult to find the entry point for the deep integration between different disciplines. These factors have led to scattered research attempts in earth system sciences. In 2009, the National Foundation of the United States recommended ten frontiers in the research field of geosciences in the twenty-first century (Geovision, NSF, 2009), but these frontiers are still mainly aimed at the research of individual branches of geosciences and are limited to the research of a single discipline. Therefore, how to establish a perfect earth system science theory is a very challenging problem.

In the field of biological evolution, the environment is usually used as a "background factor" to explore the process of life evolution. However, increasing evidence shows that biological evolution is also transforming the environment. In other words, the environment is a part of biological evolution. Life evolution is not simply adapting to the environment but developing in coordination with the environment. The coevolution of life and environment is also controlled by the deep process of the earth (Fig. 4.7).

The NCC gradually lost its original stability since the Mesozoic and reached the peak of destruction in the Early Cretaceous [38]. Geological records show that the

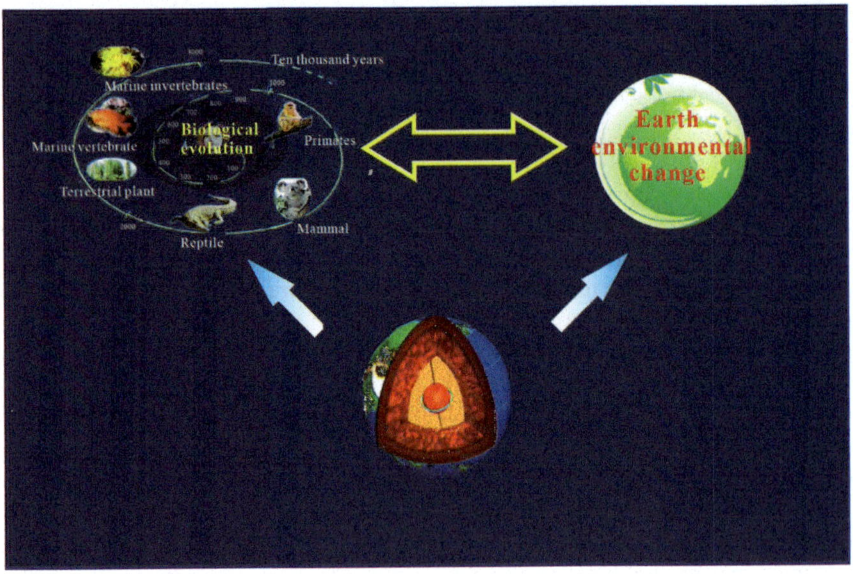

Fig. 4.7 Control of the earth's deep process on the surface system

NCC experienced three strong tectonic events from the Jurassic to the Early Cretaceous, namely, episodes A, B and C of the Yanshan Movement (Fig. 4.8). The short-term strong compressive movement caused the large-scale crustal uplift in the NCC and the formation of three regional angular unconformities. After each compressional event, the NCC was involved in large-scale volcanic eruption and crustal extension, forming widely distributed volcanic-sedimentary basins. The development and extinction of the Yanliao and Jehol biotas coincide with two major tectonic–magmatic events during the NCC destruction in time and space: the development of the Yanliao and Jehol biotas in the Late Jurassic (165–155 Ma) and Early Cretaceous (130–120 Ma) are consistent with the occurrence age of the volcanic-sedimentary basins; the extinction of the two biotas basically coincides with the time of episode B and C compressional events (Fig. 4.8). This spatiotemporal consistency indicates that there must be an internal relationship between the destruction of the NCC and the evolution of the Yanliao and Jehol biotas. The deep process of craton destruction triggered volcanic activity, crustal extension, the formation of many intermountain basins and environmental changes. These comprehensive factors may be the key factors for terrestrial biodiversity and the emergence of new species. The temporal and spatial distributions of the Yanliao and Jehol biotas also provide important information for understanding the temporal and spatial differences in the NCC destruction. The comprehensive study of the NCC destruction and the Yanliao and Jehol biotas is expected to establish a new theory on the coevolution of continental tectonics and terrestrial organisms.

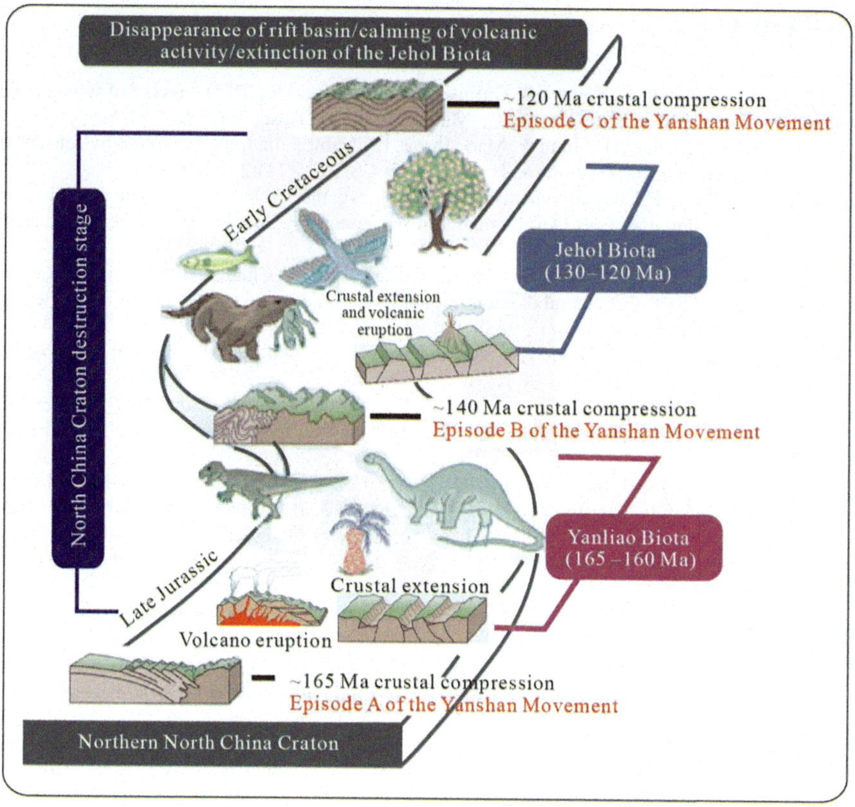

Fig. 4.8 Coupling between terrestrial biological evolution and surface system change

5 Conclusion

The implementation of the Plan has improved the international vision of Chinese geoscientists and led a group of outstanding scientists to the forefront of global geoscience. The attention of scientists in different fields extends from craton destruction to the exploration of the correlation between continental tectonics and terrestrial biological evolution, from stable continental evolution to global tectonics, from continents to oceans, and from the exploration of the deep structures of typical tectonic domains to "Deep Earth Science" to study the overall operation mechanism of the earth. This chapter can be regarded as the collective wisdom of Chinese geoscientists. We would like to express our gratitude to all colleagues who have made contributions to the Plan!

References

1. Arculus RJ, Ishizuka O, Bogus KA, Gurnis M, Hickey-Vargas R (2015) A record of spontaneous subduction initiation in the Izu–Bonin–Mariana arc. Nat Geosci 8(9):728–733
2. Chen M, Shu JF, Xie XD, Tan DY, Mao HK (2018) Natural diamond formation by self-redox of ferromagnesian carbonate. Proc Natl Acad Sci USA 115(11):2676–2680
3. Chen PJ (2000) Paleoenvironmental changes during the Cretaceous in eastern China. In: Hakuyu O, Nlall JM (eds) Developments in palaeontology and stratigraphy. Elsevier, New York, pp 81–90
4. Guilmette C, Smit MA, van Hinsbergen DJJ, Gürer D, Corfu F, Charette B, Maffione M, Rabeau O, Savard D (2018) Forced subduction initiation recorded in the sole and crust of the Semail Ophiolite of Oman. Nat Geosci 11:688–695
5. Hou ZQ, Zhang HR (2015) Geodynamics and metallogeny of the eastern Tethyan metallogenic domain. Ore Geol Rev 70:346–384
6. Hu QY, Kim DY, Yang WG, Yang LX, Meng Y, Zhang L, Mao HK (2016) FeO_2 and FeOOH under deep lower-mantle conditions and the Earth's oxygen-hydrogen cycles. Nature 534:241–244
7. Hu Q, Kim DY, Liu J, Meng Y, Yang LX, Zhang DZ, Mao WL, Mao H-K (2017) Dehydrogenation of goethite in Earth's deep lower mantle. Proc Natl Acad Sci USA 114(7):1498–1501
8. Ito M, Horikawa K (2000) Millennial- to decadal-scale fluctuation in the paleo-Kuroshio Current documented in the Middle Pleistocene shelf succession on the Boso Peninsula. Sed Geol 137(1–2):1–8
9. Jagoutz O, Macdonald FA, Royden L (2016) Low-latitude arc–continent collision as a driver for global cooling. Proc Natl Acad Sci USA 113(8):4935–4940
10. Jian Z, Yu Y, Li B, Wang J, Zhang X, Zhou Z (2006) Phased evolution of the south–north hydrographic gradient in the South China Sea since the middle Miocene. Palaeogeogr Palaeoclimatol Palaeoecol 230:251–263
11. Le Pichon X (1968) Sea-floor spreading and continental drift. J Geophys Res 73(12):3661–3697
12. Li SG, Yang W, Ke S, Meng XN, Tian HC, Xu LJ, He YS, Huang J, Wang XC, Xia QK, Sun WD, Yang XY, Ren ZY, Wei HQ, Liu YS, Meng FC, Yan J (2017) Deep carbon cycles constrained by a large-scale mantle Mg isotope anomaly in eastern China. Natl Sci Rev 4(1):111–120
13. MacKinnon A, Zhao ZX, Whalen CB (2017) Climate process team on internal wave-driven ocean mixing. Bull Am Meteor Soc 98(11):2429–2454
14. Mao HK, Hu QY, Yang LX, Liu J, Kim DY, Meng Y, Zhang L, Prakapenka VB, Yang WG, Mao WL (2017) When water meets iron at Earth's core–mantle boundary. Natl Sci Rev 4(6):870–878
15. McKenzie DP, Parker RL (1967) The North pacific: an example of tectonics on a sphere. Nature 216:1276–1280
16. Morgan WJ (1968) Rises, trenches, great faults, and crustal blocks. J Geophys Res 73(6):1959–1982
17. Müller RD, Sdrolias M, Gaina C, Steinberger B, Heine C (2008) Long term sea-level fluctuations driven by ocean basin dynamics. Science 319(5868):1357–1362
18. Müller RD, Seton M, Zahirovic S, Williams SE, Matthews KJ, Wright NM, Shephard GE, Maloney KT, Barnett-Moore N, Hosseinpour M, Bower DJ, Cannon J (2016) Ocean basin evolution and global-scale plate reorganization events since Pangea breakup. Annu Rev Earth Planet Sci 44:107–138
19. Niu YL, O'Hara MJ, Pearce JA (2003) Initiation of subduction zones as a consequence of lateral compositional buoyancy contrast within the lithosphere: a petrological perspective. J Petrol 44(5):851–866
20. Richards JP (2015) Tectonic, magmatic, and metallogenic evolution of the Tethyan orogen: from subduction to collision. Ore Geol Rev 70:323–345
21. Royden LH, Burchfiel BC, van der Hilst RD (2008) The geological evolution of the Tibetan Plateau. Science 321(5892):1054–1058
22. Sadooni FN, Alsharhan AS (2004) Stratigraphy, lithofacies distribution, and petroleum potential of the Triassic strata of the northern Arabian plate. AAPG Bull 88(4):515–538

23. Şengör AMC (1990) Plate tectonics and orogenic research after 25 years: a Tethyan perspective. Earth Sci Rev 27:1–201
24. Seton M, Müller RD, Zahirovic S, Gaina C, Torsvik T, Shephard G, Talsma A, Gurnis M, Turner M, Maus S, Chandler M (2012) Global continental and ocean basin reconstructions since 200 Ma. Earth Sci Rev 113(3–4):212–270
25. Stampfli GM, Hochard C, Vérard C, Wilhem C, von Raumer J (2013) The formation of Pangea. Tectonophysics 593:1–19
26. Stern TA, Henrys SA, Okaya D, Louie JN, Savage MK, Lamb S, Sato H, Sutherland R, Iwasaki T (2015) A seismic reflection image for the base of a tectonic plate. Nature 518:85–88
27. Stern RJ (2004) Subduction initiation: spontaneous and induced. Earth Planet Sci Lett 226(3–4):275–292
28. Sutherland R, Collot J, Bache F, Henrys S, Barker D, Browne G, Lawrence M, Morgans H, Hollis C, Clowes C, Mortimer N, Rouillard P, Gurnis M, Etienne S, Stratford W (2017) Widespread compression associated with Eocene Tonga-Kermadec subduction initiation. Geology 45(4):355–358
29. Vallis GK (2006) Atmospheric and oceanic fluid dynamics: fundamentals and large-scale circulation. Cambridge University Press, Cambridge
30. Xu YG, Li HY, Hong LB, Ma L, Ma Q, Sun MD (2018) Generation of Cenozoic intraplate basalts in the big mantle wedge under eastern Asia. Sci China Earth Sci 61(7):869–888
31. Yang JS, Robinson PT, Dilek Y (2014) Diamonds in ophiolites. Elements 10(2):127–130
32. Zheng JP, Dai HK (2018) Subduction and retreating of the western Pacific plate resulted in lithospheric mantle replacement and coupled basin–mountain respond in the North China Craton. Sci China Earth Sci 61(4):406–424
33. Zheng YF, Xu Z, Zhao ZF, Dai LQ (2018) Mesozoic mafic magmatism in North China: implications for thinning and destruction of cratonic lithosphere. Sci China Earth Sci 61(4):353–385
34. Zhou Z (2014) The Jehol Biota, an Early Cretaceous terrestrial Lagerstätte: new discoveries and implications. Natl Sci Rev 1(4):543–559
35. Zhu RX, Chen L, Wu FY, Liu JL (2011) Timing, scale and mechanism of the destruction of the North China Craton. Sci China Earth Sci 54(6):789–797
36. Zhu RX, Xu YG, Zhu G, Zhang HF, Xia QK, Zheng TY (2012) Destruction of the North China Craton. Sci China Earth Sci 55(10):1565–1587
37. Zhu RX, Fan HR, Li JW, Meng QR, Li SR, Zeng QD (2015) Decratonic gold deposits. Sci China Earth Sci 58(9):1523–1537
38. Zhu RX, Zhang HF, Zhu G, Meng QR, Fan HR, Yang JH, Wu FY, Zhang ZY, Zheng TY (2017) Craton destruction and related resources. Int J Earth Sci 106:2233–2257

The manufacturer's authorised representative in the EU is Springer Nature Customer Service Centre GmbH, Europaplatz 3, 69115 Heidelberg, Germany. If you have any concerns regarding our products, please contact ProductSafety@springernature.com

Printed and bound by CPI Group (UK) Ltd, Croydon, CR0 4YY
01/10/2026
02225883-0001